PROPERTIES OF THE THIRTY-TWO POINT GROUPS

PROPERTIES OF THE THIRTY-TWO POINT GROUPS

PUBLISHED 1963 BY THE M.I.T. PRESS, CAMBRIDGE, MASSACHUSETTS

GEORGE F. KOSTER

JOHN O. DIMMOCK

ROBERT G. WHEELER

HERMANN STATZ

Second printing, December 1966

Library of Congress Catalog Card Number: 63-21194
Printed in the United States of America

FOREWORD

There has long been a need in science and engineering for systematic publication of research studies larger in scope than a journal article but less ambitious than a finished book. Much valuable work of this kind is now published only in a semiprivate way, perhaps as a laboratory report, and so may not find its proper place in the literature of the field. The present contribution is the twenty-fourth of the M.I.T. Press Research Monographs, which we hope will make selected timely and important research studies readily accessible to libraries and to the independent worker.

J. A. Stratton

CONTENTS

Part 1

DISCUSSION

Part 2

TABLES

List of Figures

Part 1

DISCUSSION

Chapter 1

INTRODUCTION

For a discussion of the quantum theory of the energy levels of atoms perturbed by a crystalline environment or for a discussion of the energy levels associated with the periodic potential problem, it is essential to know the irreducible representations of the thirty-two crystallographic point groups and their associated double groups. Character tables for the crystallographic point groups exist in a number of references.[1-5] Those for the associated double groups were, until recently, scattered in a number of separate articles.[1,5-9] Koster,[10] in an article reviewing the irreducible representations of the space groups, has given the character tables for all of the double groups associated with the point groups. It would seem therefore that there is little point in presenting again these character tables. Unfortunately this is not the case for two reasons. The first reason comes about as follows: A standard situation that occurs in the consideration of energy levels in a crystal is the discussion of a succession of perturbations that gradually reduce the symmetry of the environment. Thus one might discuss first the effect of a cubic perturbation on atomic eigenstates and then discuss the further splitting induced by a perturbation of tetragonal symmetry. In another case, one might want to discuss the effect of a magnetic field on the atom in some crystalline environment. In order to do this, it is necessary to consider the larger group of symmetries and a succession of subgroups corresponding to the reduction in symmetry caused by successive perturbations. In order to carry out such a discussion, it is convenient to have the notation for the operators be consistent as we pass from group to subgroup. Unfortunately, the complete tabulation of Koster[10] suffers from the fact that there are awkward shifts in notation in the labeling of operations of a group and one of its possible subgroups. (In addition, the characters for the complex double group representations of the group T in this compilation[10] are wrong, the characters for the operation C_3 and $\bar{C}_3$ being interchanged for these two complex representations.)

The second reason for reproducing the character tables is that we want to discuss the coupling coefficients for the crystallographic point groups. (These will be discussed in detail in Section 3.2.) The coupling coefficients are, for the point groups, the analogue of the Clebsch-Gordan coefficients for the full rotation group and are useful in the same way that the latter coefficients find application in the discussion of the quantum theory of spherical systems. The coupling coefficients depend on the choice of basis for the irreducible representations of the point groups. Before the tables can be given, the bases and the irreducible representations to which they correspond must be given. In order to do this, the character tables must be reproduced along with a notation for these representations.

For these reasons it seems appropriate to collect in one place the character tables of all of the crystallographic point and double groups once again. With these we shall present tables of the coupling coefficients and supplementary tables that enable us to thread our way through a succession of subgroups that are important when a sequence of perturbations enters into the discussion of energy levels in crystalline fields with applied magnetic and electric fields. In the next chapter, we discuss in detail the notation and conventions that we have used in the definition of the operators for the crystallographic point groups and their associated double groups.

Chapter 2

NOTATION AND MULTIPLICATION CONVENTIONS FOR THE GROUP OPERATORS

For the operators of the thirty-two crystallographic point groups we shall use the following standard notation:

E = identity
C_6 = rotation through +60°
C_3 = rotation through +120°
C_4 = rotation through +90°
C_2 = rotation through +180°

It is a group requirement that the inverses R^{-1} (where $RR^{-1} = R^{-1}R = E$) of all operators R of the group must also be members of the group, such that we have also

C_6^{-1} = rotation through −60°
C_3^{-1} = rotation through −120°
C_4^{-1} = rotation through −90°
C_2^{-1} = rotation through −180°

In addition to these pure rotations we have the following improper rotations:

I = inversion
$\sigma = IC_2 = C_2I$ = a reflection in a plane perpendicular to the axis of rotation of C_2;
σ_h is a reflection in a plane perpendicular to a principal axis of symmetry;
σ_v is a reflection in a plane containing a principal axis of symmetry;
σ_d is a reflection containing a principal axis of symmetry that bisects the angle between two twofold rotation axes perpendicular to the principal axis;
S_n ($n = 3$, $n = 4$, and $n = 6$) will denote a rotation C_n followed by a reflection in a plane perpendicular to the axis of rotation.

The operators, which we have defined earlier, can be thought of as operating on scalar functions of position; they have the effect of rotating these functions of position (we shall define this more carefully in Chapter 3 when discussing the coupling coefficients). In problems involving the spin of the electron, we must include in our operators a factor which describes the way a spinor behaves under the physical rotation (proper or improper) that we have in mind.[11] In this way, as is well known, corresponding to each rotation operator that operates on scalar functions of position, we get two operators, one being the negative of the other. Corresponding to the identity there will be two operators that we shall call E and $\overline{E}$. The operator $\overline{E}$ is identical to E except that it changes the sign of the spinor. For the other operators we shall use the notation R and $\overline{R}$ for the two operators that correspond to the same physical rotation of a set of coordinate axes. The two operators differ only in that one changes the sign of the spinor relative to the other ($\overline{R} = \overline{E}R$). When considering these operators and the groups that can be constructed from them, we are led to the familiar double groups corresponding to the thirty-two crystallographic point groups. These groups can be divided into classes just as the ordinary crystallographic groups are divided. The following rule for passing from the classes of the latter groups to the classes of the double groups was obtained by Opechowski:[12]

> The Class of R and the class of $\overline{R}$ are different for all proper rotations except those through 180°. An unbarred class of twofold rotations is different from the barred class unless there is a twofold rotation about an axis perpendicular to the original twofold rotation or a reflection in a plane containing the axis. In either of these cases, the barred and the unbarred classes combine into a single class of the double group. The situation is the same for the improper rotations. In this case the class of barred operations and the corresponding class of unbarred operations are different except in the case of reflections. Then a barred and an unbarred reflection are in different classes unless there is either a reflection in a plane perpendicular to the given reflection plane or a twofold rotation about an axis in the plane of reflection.

Consider now the multiplication of barred and unbarred pure rotations. The situation may be visualized by considering that by a rotation through 360° we do not return to the original state but obtain a change in the sign of the spinor.[11] This operation is thus $\overline{E}$. An additional rotation through 360° brings us back to the identity E. The multiplication table for rotations can be represented diagramatically in such a way that a 360° rotation of the diagram represents a physical rotation of 720°. The rotation multiplication table set up in this way is illustrated in Figure 1. In the figure we have preserved the relationship $RR^{-1} = E$ for the barred operations.

We must now give the rules for multiplications involving improper rotations. The inversion operator I has the quality that it commutes with all point operators, leaves spinor functions invariant, and

$$II = E \qquad I\overline{I} = \overline{E} \qquad \overline{I}\,\overline{I} = E$$

Hence I times an unbarred operator gives an unbarred operator, and I times a barred operator gives a barred operator. That is,

$$IR_i = S_j \qquad I\overline{R}_i = \overline{S}_j \quad \text{etc.}$$

If R_i, R_j, and C_2 are pure rotations about the same axis, we have from Figure 1,

$$C_2 R_i = \begin{cases} \bar{R}_j & \text{if } 0^\circ < R_i < 180^\circ; \\ R_j & \text{if } -180^\circ < R_i \leq 0^\circ; \end{cases}$$

Because of the properties of the inversion we have

$$IC_2 = \sigma_h$$

where σ_h is a reflection in a plane perpendicular to the axis of rotation of C_2.

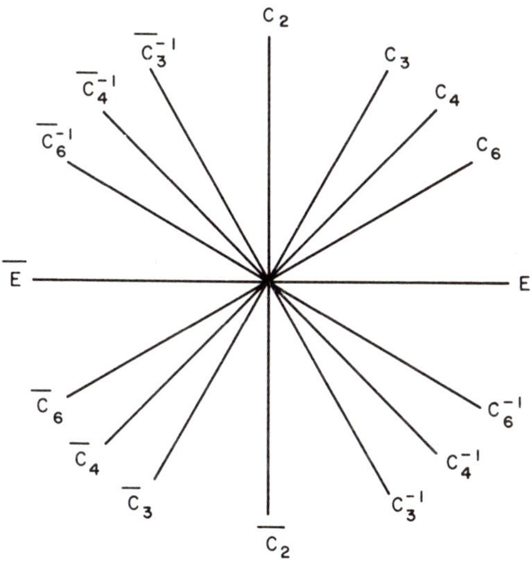

Figure 1. Schematic multiplication of double group rotation operators.

We have the following consequences:

$$\sigma_h \sigma_h = IC_2 (IC_2) = C_2 C_2 = \bar{E}$$

since I commutes with all operators. From this we see that

$$\sigma_h \bar{\sigma}_h = E$$

or that $\sigma_h^{-1} = \bar{\sigma}_h$. These relations are true for all of the reflections no matter what the plane of reflection. From the previous relations we easily conclude that

$$\sigma_h R_i = \begin{cases} \bar{S}_i & \text{if } 0^\circ < R_i < 180^\circ; \\ S_i & \text{if } -180^\circ < R_i < 0^\circ; \end{cases}$$

We must now examine the operation of a rotation times a reflection in a plane containing the axis of rotation. Consider the group C_{3v} for example. This consists of 6 classes:

$$E$$
$$\bar{E}$$
$$(C_3, C_3^{-1})$$
$$(\bar{C}_3, \bar{C}_3^{-1})$$
$$3\sigma_v$$
$$3\bar{\sigma}_v$$

We have two possibilities:

$$C_3\sigma_v = \sigma_v' \qquad \text{or} \qquad C_3\sigma_v = \bar{\sigma}_v'$$

In either of these cases we have

$$C_3C_3\sigma_v = \sigma_v''$$

but

$$C_3C_3C_3\sigma_v = \bar{E}\sigma_v = \bar{\sigma}_v$$

whence

$$C_3\sigma_v'' = \bar{\sigma}_v$$

which then establishes in general that

$$C_3\sigma_v = \bar{\sigma}_v, \quad \text{etc.}$$

similarly,

$$C_3^{-1}\sigma_v = \bar{\sigma}_v'', \quad \text{etc.}$$

and consequently,

$$\sigma_v''\sigma_v' = \sigma_v\sigma_v'' = \sigma_v'\sigma_v = C_3$$

$$\sigma_v'\sigma_v'' = \sigma_v''\sigma_v = \sigma_v\sigma_v' = C_3^{-1}$$

The remaining multiplication relations that cannot be determined from those cited here are free to be defined as we wish provided the definitions are consistent with each other and with the multiplications already obtained. The character tables to be presented are not affected by these free definitions and are thus completely determined by the cited considerations. These character tables will be presented along with the coupling coefficients for the thirty-two groups. These coefficients will be discussed in Chapter 3.

Chapter 3

COUPLING COEFFICIENTS

3.1 General Discussion

A standard problem encountered in the quantum theory of angular momentum is the coupling of the magnetic sublevels of two angular momentum operators $\mathbf{J}_1$ and $\mathbf{J}_2$ to give eigenstates of the total angular momentum $\mathbf{J} = \mathbf{J}_1 + \mathbf{J}_2$ and the z component of this angular momentum. This is done through the use of the Clebsch-Gordan coefficients as follows: Imagine we call $u(J_1,m_1)$ and $v(J_2,m_2)$ eigenstates of the angular momentum operators $\mathbf{J}_1$ and $\mathbf{J}_2$ such that

$$\begin{aligned}
\mathbf{J}_1^2 u(J_1,m_1) &= J_1(J_1+1)u(J_1,m_1) \\
\mathbf{J}_{1z} u(J_1,m_1) &= m_1 u(J_1,m_1) \\
\mathbf{J}_2^2 v(J_2,m_2) &= J_2(J_2+1)v(J_2,m_2) \\
\mathbf{J}_{2z} v(J_2,m_2) &= m_2 v(J_2,m_2)
\end{aligned} \tag{3.1}$$

Here the index z denotes the z component of the appropriate angular momentum vector. We now want to find those linear combinations of the $(2J_1+1)(2J_2+1)$ product states $u(J_1,m_1)v(J_2,m_2)$ (where $m_1 = -J_1 \ldots J_1$; $m_2 = -J_2 \ldots J_2$) such that these combinations are eigenstates of the total angular momentum $\mathbf{J}^2$ and its z component $\mathbf{J}_z$. The coefficients in this linear combinations are called the Clebsch-Gordan coefficients $c(J,m;\ J_1m_1,J_2m_2)$.[12]

$$\left.\begin{aligned}
&\psi(J,m) = \sum_{(m_1,m_2)} c(Jm;J_1m_1,J_2m_2)u(J_1m_1)v(J_2m_2) \\
\text{where}\quad &\mathbf{J}^2\psi(J,m) = J(J+1)\psi(J,m) \\
&\mathbf{J}_z\psi(J,m) = m\psi(J,m)
\end{aligned}\right\} \tag{3.2}$$

We can phrase this procedure in a different way. It is well-known that the functions $u(J_1m_1)$ (where $m_1 = -J_1 \cdots J_1$) form a basis for a $2J_1+1$ dimensional irreducible representation of the full three-dimensional rotation group. The product functions $u(J_1m_1)v(J_2m_2)$ form a basis for a $(2J_1+1)(2J_2+1)$ dimensional representation of the full rotation group. This representation is, however, in general reducible.[13] We might ask to find the matrix that reduces this direct product representation of the full rotation group. It can easily be seen that the columns of the matrix that reduces this representation are nothing more than the Clebsch-Gordan coefficients $c(Jm;J_1m_1,J_2m_2)$ arranged in dictionary order on the indices m_1 and m_2. The matrix formed out of these columns by labeling the columns by the indices J, m (where $J = |J_1 - J_z| \cdots J_1 + J_2$; $m = -J \cdots J$) is just this matrix that reduces the direct product representation. When viewed in this way, it is clear that there is a generalization that can be made. In the present case we are concerned not with full spherical symmetry, but rather with symmetries corresponding to the crystallographic point groups. We can ask for those matrices which reduce direct products of the irreducible representations of these groups and then use the columns of these matrices to couple product states to give resultant states that again transform irreducibly. In the tables in Part 2, we give all of the coupling coefficients that reduce all possible direct products of representations of the crystallographic point groups. In the discussion, we give the notation and conventions used in the construction of these tables.

3.2 Definition of the Coupling Coefficients

We imagine that we have functions $u_i^\alpha (i = 1 \cdots n_\alpha)$ and v_j^β $(j = 1 \cdots n_\beta)$ which transform irreducibly under some group G of operators. The dimensions of the irreducible representations are n_α and n_β. We call the matrices in these irreducible representations $\underset{\sim}{\Gamma}_\alpha(R)$, where R is some operator of the group G. Thus we have

$$\left.\begin{aligned} Ru_i^\alpha &= \Sigma(k)\Gamma_\alpha(R)_{ki}u_k^\alpha \\ Rv_j^\beta &= \Sigma(m)\Gamma_\beta(R)_{mj}v_m^\beta \end{aligned}\right\} \tag{3.3}$$

The $n_\alpha n_\beta$ product functions $u_i^\alpha v_j^\beta$ form a basis for the reducible representation $\underset{\sim}{\Gamma}(R)$. The matrix elements in this representation can easily be seen to be

$$\Gamma(R)_{(km)(ij)} = \Gamma_\alpha(R)_{ki}\Gamma_\beta(R)_{mj} \tag{3.4}$$

where the indices specifying the columns and rows of the matrix $\underset{\sim}{\Gamma}(R)$ can be taken to be the pairs (ij) and (km), respectively, taken in dictionary order. We now want to find those linear combinations of products which themselves transform irreducibly under G. In other words, we wish to find coefficients $U_{ij,n}^{\alpha\beta,\gamma}$ such that the linear combinations

$$\psi_n^\gamma = \Sigma(i,j)U_{ij,n}^{\alpha\beta,\gamma}\, u_i^\alpha v_j^\beta \tag{3.5}$$

have the property that

$$R\psi_n^\gamma = \Sigma(p)\Gamma_\gamma(R)_{pn}\,\psi_p^\gamma \tag{3.6}$$

where $\underset{\sim}{\Gamma}_\gamma(R)$ is an irreducible representation of the group G. These coefficients $U_{ij,n}^{\alpha\beta,\gamma}$ are what we call the coupling coefficients. Whether or not we can find linear combinations transforming according to the nth row of the γ irreducible representation depends on the number of times this representation is contained in the decomposition of the direct product representation (Equation 3.4) into irreducible representations. It is a well-known result of group theory that this number is given by

$$c_{\alpha\beta\gamma} = \frac{1}{g}\ \Sigma(R)\chi_\alpha(R)\chi_\beta(R)\chi^*_\gamma(R) \tag{3.7}$$

here $\chi(R)$ is the character of a representation and g is the order of the group; $\chi_\alpha(R)$ is the character of the αth representation of the operator R:

$$\chi_\alpha(R) = \Sigma(i)\Gamma_\alpha(R)_{ii} \tag{3.8}$$

If $c_{\alpha\beta\gamma}$ is zero, then we cannot find any linear combination of the product functions transforming according to the irreducible representation $\underset{\sim}{\Gamma}_\gamma(R)$. If $c_{\alpha\beta\gamma} = 1, 2$, etc., then we can find 1,2, etc., functions transforming according to a given row of the γ representation. It is at this point that there is a difference between the point groups and the full rotation group. For the full rotation group a given representation (given J) is either contained once or not at all in the decomposition of the direct product of the irreducible representations corresponding to the angular momenta J_1 and J_2. In the case of the point groups, a representation can be contained once, twice, or not at all in the decomposition of the direct product.‡ Let us now relate the coefficients $U^{\alpha\beta,\gamma}_{ij,n}$ to the matrix reducing $\underset{\sim}{\Gamma}(R)$. It can be easily seen by operating on Equation 3.5 with R and comparing the result with the right side of Equation 3.6, after replacing the ψ's on the right side with Equation 3.5, that the matrix $\mathbf{U}^{\alpha\beta}$ with the pair γn labeling columns and the pair ij labeling rows is just the matrix which puts $\underset{\sim}{\Gamma}(R)$ in reduced form. Thus the matrix

$$(\mathbf{U}^{\alpha\beta})^\dagger \underset{\sim}{\Gamma}(R)\mathbf{U}^{\alpha\beta} \tag{3.9}$$

is in reduced form with the irreducible representations occurring in the reduction appearing in blocks down the main diagonal. (The symbol † indicates the complex conjugate transpose.) This matrix will be unitary since we demand that the u's, v's, and ψ's are normalized. The coupling coefficients listed in the tables in Part 2 are nothing more than the **U**'s defined earlier. It should be clear that in order to define uniquely the coupling coefficients we must first define the irreducible representations of the point groups. In the next section (Section 3.3), we discuss the conventions used in this definition.

3.3. Time Reversal and Conventions

The discussion of the last paragraph of Section 3.2 is applicable to any finite group. Here we are concerned with the thirty-two crystallographic point groups. We deal with the operators of these groups, which are real orthogonal coordinate transformations, in the following way: Imagine we have a scalar function of position f(x,y,z) and a real orthogonal coordinate transformation

‡It is clear, at this point, that we should have another index on the U's and the ψ's. This index would specify which of the linear combinations transform according to the γ representation. For simplicity in notation we shall not introduce this additional index.

$$\left.\begin{aligned} x' &= R_{11}x + R_{12}y + R_{13}z \\ y' &= R_{21}x + R_{22}y + R_{23}z \\ z' &= R_{31}x + R_{32}y + R_{33}z \end{aligned}\right\} \tag{3.10}$$

Let us define an operator R so that

$$f(x',y',z') = R^{-1}f(x,y,z) \tag{3.11}$$

Any real orthogonal coordinate transformation can be considered as either proper (a rotation about some axis through an angle θ) or improper (a rotation through an angle θ about some axis followed by a reflection through a plane perpendicular to that axis). The convention we use is the following: If we take the z axis as the axis of rotation, the rotation is clockwise about the +z direction, the +x direction being rotated into the +y direction through an angle $+\theta$. The matrix $[R_{ij}]$ for this operation is

$$\begin{bmatrix} \cos\theta & \sin\theta & 0 \\ -\sin\theta & \cos\theta & 0 \\ 0 & 0 & \pm 1 \end{bmatrix} \tag{3.12}$$

The + gives the proper rotation that we have described. The − sign gives the same proper rotation followed by a reflection through the xy plane (an improper rotation). The thirty-two crystallographic point groups consist of such operations, performed with a point held fixed, which send a periodic lattice into itself. As we mentioned in an earlier section, we are also concerned with the double groups associated with these crystallographic groups. This complication arises in quantum mechanical problems involving the spin of the electron. In these cases we must deal with spinor rather than scalar functions of position. We must consider these double groups in a little more detail in order to explain the conventions used in our tabulation of coupling coefficients.

The groups of operators that transform a two component spinor function of position

$$\Phi = (\phi_{-\frac{1}{2}}, \phi_{\frac{1}{2}}) \tag{3.13}$$

contain operators whose form is

$$\pm \mathbf{D}_{\frac{1}{2}}(R)R \tag{3.14}$$

where

$$\left.\begin{aligned} D_{-\frac{1}{2},-\frac{1}{2}}(R) &= D_{\frac{1}{2},\frac{1}{2}}(R)^* = \cos\frac{\beta}{2}\exp\left[\left(-\frac{i}{2}\right)(\alpha+\gamma)\right] \\ D_{-\frac{1}{2},\frac{1}{2}}(R) &= -D_{\frac{1}{2},-\frac{1}{2}}(R)^* = \sin\frac{\beta}{2}\exp\left[\left(-\frac{i}{2}\right)(\alpha-\gamma)\right] \end{aligned}\right\} \tag{3.15}$$

Here the operator R corresponds to a coordinate transformation described by

Euler angles α,β,γ.[14] These correspond to the following transformation of our coordinate system: (a) a rotation of the coordinates about the z axis by an angle γ, (b) a rotation of the coordinates about the new y axis by an angle β, and (c) a rotation of the coordinates about the new z axis by an angle α. When applied to a spinor function of position $\underset{\sim}{\Phi}$, these operators give

$$\underset{\sim}{\Phi}' = \pm(R\phi_{-\frac{1}{2}},\ R\phi_{\frac{1}{2}})\mathbf{D}_{\frac{1}{2}}(R) \tag{3.16}$$

Because of the well-known $\pm$ sign ambiguity, we get, corresponding to each of the thirty-two crystallographic point groups, a group of twice the order: one of the crystallographic double point groups. These double point groups are the groups which are considered in this discussion.

In order to define uniquely the coupling coefficients for these groups, we must specify the irreducible representations precisely. This we do in terms of the bases for these representations (Equation 3.3). In the character tables of Part 2, we give, for each representation of a particular group, the basis functions used in the definition of that representation along with the order in which the basis functions occur in the case of a multidimension representation. (The notation will be discussed in Section 4.1).

There is another point that must be discussed in connection with the coupling coefficients. Often in quantum mechanical problems we consider, in addition to the spatial operators (Equation 3.14), the operator of time inversion. This operator, which we shall denote by $\mathbf{K}$, has been discussed by Wigner.[15] It has the effect, when applied to a Hamiltonian $\mathbf{H}$, of giving a transformed Hamiltonian $\mathbf{K}^{-1}\mathbf{H}\mathbf{K}$ in which every coordinate is replaced by itself and every momentum by its negative. It also has the property that $\mathbf{K}^2 = E$ for an even number of electrons and $\mathbf{K}^2 = -E$ for an odd number of electrons. It can be shown that for each electron this operator has the form of a two-by-two unitary matrix multiplying the operator which converts every function into its complex conjugate. The connection between this operator and the discussions of the coupling coefficients for irreducible representations presented here is as follows:

Imagine that we have a Hamiltonian $\mathbf{H}$ which is invariant under time inversion, and a set of degenerate eigenstates $u_1 \cdots u_n$. We assume that these states form the basis for an irreducible representation $\underset{\sim}{\Gamma}(R)$ of the group of unitary operators which commute with the Hamiltonian. We know then that the set of functions $\mathbf{K}u_1 \cdots \mathbf{K}u_n$ have the same energy as the set $u_1 \cdots u_n$. It is easy to show that the set of functions $\mathbf{K}u_i$ form a basis for the representation $\underset{\sim}{\Gamma}(R)^*$. There are three possible relations between a representation and its complex conjugate.[15] We call these three cases a, b, and c.

a. $\underset{\sim}{\Gamma}(R) = \underset{\sim}{\Gamma}(R)^*$; representation real.

b. $\underset{\sim}{\Gamma}(R)$ complex and not equivalent to $\underset{\sim}{\Gamma}(R)^*$.

c. $\underset{\sim}{\Gamma}(R)$ complex and equivalent to $\underset{\sim}{\Gamma}(R)^*$ but cannot be made real.

With this classification Wigner was able to show that six relationships are possible between the states u_i^α and $\mathbf{K}u_i^\alpha$ depending on whether the number of electrons is even or odd and depending on whether we have case a, b, or c.

Even numbers of electrons ($\mathbf{K}^2 = E$):

a. The set $\mathbf{K}u_i^\alpha$ can be chosen in such a way that $\mathbf{K}u_i^\alpha = u_i^\alpha$, and we have no degeneracy in addition to the n-fold degeneracy.
b. The set $\mathbf{K}u_i^\alpha$ is linearly independent of the set u_i^α. The degeneracy is now 2n; u_i is a basis for $\underset{\sim}{\Gamma}_\alpha(R)$ and $\mathbf{K}u_i^\alpha$ is a basis for $\underset{\sim}{\Gamma}_\alpha(R)^* = \underset{\sim}{\Gamma}_\beta(R)$.
c. This case does not occur in the crystallographic point groups.

Odd numbers of electrons ($\mathbf{K}^2 = -E$):

a. The set $\mathbf{K}u_i^\alpha$ is linearly independent of the set u_i^α. The degeneracy is now 2n. Both the set u_i^α and the set $\mathbf{K}u_i^\alpha$ form a basis for the same representation $\underset{\sim}{\Gamma}_\alpha(R)$.
b. The set $\mathbf{K}u_i^\alpha$ is linearly independent of the set u_i^α. The degeneracy is 2n. The set u_i^α is a basis for $\underset{\sim}{\Gamma}_\alpha(R)$ and the set $\mathbf{K}u_i^\alpha$ for $\underset{\sim}{\Gamma}_\alpha(R)^* = \underset{\sim}{\Gamma}_\beta(R)$.
c. The set $\mathbf{K}u_i^\alpha$ is linearly related to the set u_i^α, and we have no degeneracy in addition to the n-fold degeneracy. Wigner[15] has shown that in this case n must be even.

In general in the case of an odd number of electrons we can label the basis functions with the subscript i such that $i = -q \cdots -1/2, 1/2 \cdots + q$; i taking on half integral values from $-q$ up to $+q$. (1/2,3/2, etc.). The dimensionality of the representation is $2q + 1$. The linear relation between u_i^α and $\mathbf{K}u_i^\alpha$ can be taken in the form[16]

$$\mathbf{K}u_i = (-1)^{q-i}u_{-i} \tag{3.17}$$

From the characters of a representation we can determine whether the representation is of type a, b, or c. In the character tables, we signify by these letters whether a representation belongs to case a, b, or c. The basis functions have been chosen such that they follow the conventions listed earlier. Therefore, we demand that the linear combinations of product functions, formed through the coupling coefficients, also satisfy our time reversal conventions. The way in which this influences our results can be seen from a simple example.

Consider the case of the double group corresponding to the point group that consists of just the identity spatial operator. This is a group of order two consisting of E and $\bar{E}$. This group has a symmetric representation Γ_1 and an antisymmetric representation $\underset{\sim}{\Gamma}_2$. Consider a function u and a function v that transform according to $\underset{\sim}{\Gamma}_2$ (odd number of electrons, case a). (See the table for the group C_1.) It is clear that under the two operators of this group the product of these two functions transforms like $\underset{\sim}{\Gamma}_1$. From our definition of the coupling coefficients it would appear that there is just one coupling coefficient whose value is 1. Let us consider, however, our time reversal conventions. Both $\mathbf{K}u$ and $\mathbf{K}v$ are linearly independent of u and v, respectively, and also, separately, each forms a basis for $\underset{\sim}{\Gamma}_2$. From this we see that

$\mathbf{K}u(\mathbf{K}v)$ is not equal to uv. From our conventions (case a for $\underset{\sim}{\Gamma}_1$, even number of electrons) we see that we must find linear combinations that go into themselves under time reversal. Remembering that $\mathbf{K}^2 = -E$, we can easily see that the combinations

$$\frac{1}{\sqrt{2}}\ [uv + (\mathbf{K}u)\mathbf{K}v]$$

$$\frac{i}{\sqrt{2}}\ [uv - (\mathbf{K}u)\mathbf{K}v]$$

$$\frac{1}{\sqrt{2}}\ [u\mathbf{K}v - (\mathbf{K}u)v]$$

$$\frac{i}{\sqrt{2}}\ [u\mathbf{K}v + (\mathbf{K}u)v]$$

have the desired property. These, therefore, are the coupling coefficients that will be found in the table under the group C_1.

3.4. Applications of the Coupling Coefficients

In addition to the usefulness in finding those linear combinations of product functions that transform irreducibly under a given group, there is another important application of the coupling coefficients. They are useful in the computation of the matrix elements of operators that transform irreducibly under a given group.[17] Suppose we had a set of n_α operators $\mathbf{P}_i^\alpha$ that transform irreducibly under some group:

$$R\mathbf{P}_i^\alpha R^{-1} = \Sigma(k)\Gamma_\alpha(R)_{ki}\mathbf{P}_k^\alpha \tag{3.18}$$

Let us take the matrix elements of this operator between states u_n^γ and $v_j^\beta (n = 1 \cdots n_\gamma,\ j = 1 \cdots n_\beta)$, where we assume that these functions form bases for the representations $\underset{\sim}{\Gamma}_\gamma(R)$ and $\underset{\sim}{\Gamma}_\beta(R)$, respectively. These $n_\alpha n_\beta n_\gamma$ matrix elements

$$\text{where}\quad \left.\begin{array}{c} (u_n^\gamma \mid \mathbf{P}_i^\alpha \mid v_j^\beta) \\ n = 1 \ldots n_\gamma \\ i = 1 \ldots n_\alpha \\ j = 1 \ldots n_\beta \end{array}\right\} \tag{3.19}$$

can be expressed in terms of as many independent constants as the number of times the representation $\underset{\sim}{\Gamma}_\alpha \times \underset{\sim}{\Gamma}_\beta$ contains the representation $\underset{\sim}{\Gamma}_\gamma$, that is, $c_{\alpha\beta\gamma}$). The relations between the matrix elements are given in terms of the unitary matrix reducing the direct product of the α and β irreducible representations. This can most easily be seen as follows:[18] It is clear that the $n_\alpha n_\beta$ functions $\mathbf{P}_i^\alpha v_j^\beta$ transform according to the representation $\underset{\sim}{\Gamma}_\alpha(R)\times\underset{\sim}{\Gamma}_\beta(R)$:

$$\mathbf{R}\mathbf{P}_i^\alpha v_j^\beta = \mathbf{R}\mathbf{P}_i^\alpha R^{-1} R v_j^\beta$$

$$= \Sigma(k,m)\Gamma_\alpha(R)_{ki}\Gamma_\beta(R)_{mj}\mathbf{P}_k^\alpha v_m^\beta \tag{3.20}$$

and that the linear combinations

$$w_{n'}^{\gamma'} = \Sigma(i,j)U_{ij,n'}^{\alpha\beta,\gamma'}\mathbf{P}_i^\alpha v_j^\beta \tag{3.21}$$

transform according to the γ' irreducible represenation. (We recall at this point that there will be $c_{\alpha\beta\gamma'}$ such linear combinations which transform in precisely the same way.) Remembering that the U's are elements of a unitary matrix, we see that

$$\Sigma(\gamma',n')U_{ij,n'}^{\alpha\beta,\gamma'*} w_{n'}^{\gamma'} = \mathbf{P}_i^\alpha v_j^\beta \tag{3.22}$$

Let us now insert Equation 3.22 into Equation 3.19 and remember that

$$(u_n^\gamma \mid w_{n'}^{\gamma'}) = \delta_{\gamma\gamma'}\delta_{nn'} \text{x constant} \tag{3.23}$$

where the constant we have in Equation 3.23 is independent of n. We see that

$$\left.\begin{array}{l}
\text{if } c_{\alpha\beta\gamma} = 0 \\
\qquad (u_n^\gamma \mid \mathbf{P}_i^\alpha \mid v_j^\beta) = 0 \\
\text{if } c_{\alpha\beta\gamma} = 1 \\
\qquad (u_n^\gamma \mid \mathbf{P}_i^\alpha \mid v_j^\beta) = U_{ij,n}^{\alpha\beta,\gamma*} \times \text{(constant independent} \\
\qquad\qquad \text{of i, j, and n)} \\
\text{if } c_{\alpha\beta\gamma} = 2 \\
\qquad (u_n^\gamma \mid \mathbf{P}_i^\alpha \mid v_j^\beta) = c_1 U_{ij,n}^{\alpha\beta,\gamma*} + c_2 U_{ij,n}'^{\alpha\beta,\gamma*}
\end{array}\right\} \tag{3.24}$$

where c_1 and c_2 are both constants independent of i,j, and n. Two **U**'s appear in the relationship when $c_{\alpha\beta\gamma} = 2$ since in this case there are two linear combinations of product functions (two w's) which transform according to the nth row of the γth irreducible representation. One linear combination involves what we have called the unprimed **U**'s, and the second linear combination involves the primed **U**'s. From this argument we see that the relations between the matrix elements (Equation 3.19) are contained in the **U** matrices which reduce the direct product representations (Equation 3.11). In a later section, after discussing the form of the tables, we shall illustrate the use of the coefficients in the calculation of the matrix elements of symmetric operators.

Chapter 4

DISCUSSION OF TABLES

In this compendium all the information for each set of isomorphic point groups follows sequentially a description of the groups. The tabular material consists of the character table and basis functions, the multiplication tables showing how the direct products of the irreducible representations decompose, the tables of coupling coefficients, the compatibility tables between the groups and their subgroups, and the compatibility tables between the full spherical group and the point groups. All of these tables are discussed in detail in this chapter. Table 1 gives the designations for the thirty-two point groups in both the International form and the Schönflies notation.[19]

The tables for the groups C_2, C_s, C_{2h}, C_{2v}, D_2, D_{2h}, C_4, S_4, C_{4h}, D_4, C_{4v}, D_{2d}, and D_{4h} follow Table 1. Figure 2 serves as an illustration for these groups.

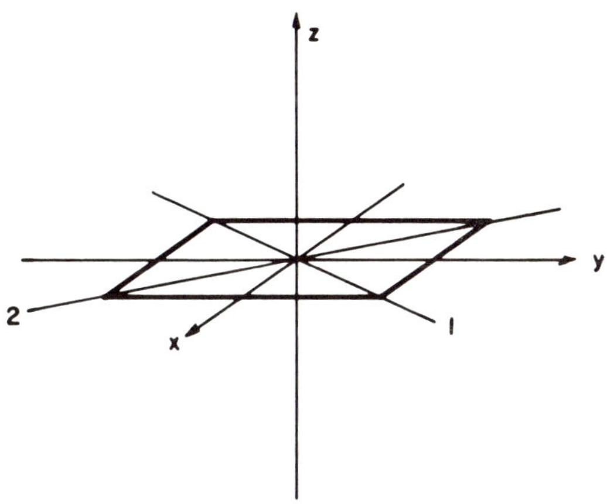

Figure 2. An illustration for the groups C_2, C_s, C_{2h}, C_{2v}, D_2, D_{2h}, C_4, S_4, C_{4h}, D_4, C_{4v}, D_{2d}, and D_{4h}.

Figure 3 serves as an illustration for the groups C_3, C_{3i}, D_3, C_{3v}, D_{3d}, C_6, C_{3h}, C_{6h}, D_6, C_{6v}, D_{3h}, and D_{6h}. Figure 4 serves to illustrate the groups T, T_h, T_d, O, and O_h.

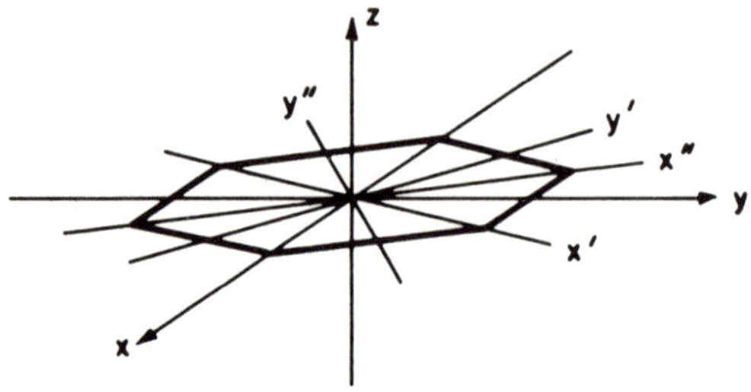

Figure 3. An illustration for the groups C_3, C_{3i}, D_3, C_{3v}, D_{3d}, C_6, C_{3h}, C_{6h}, D_6, C_{6v}, D_{3h}, and D_{6h}.

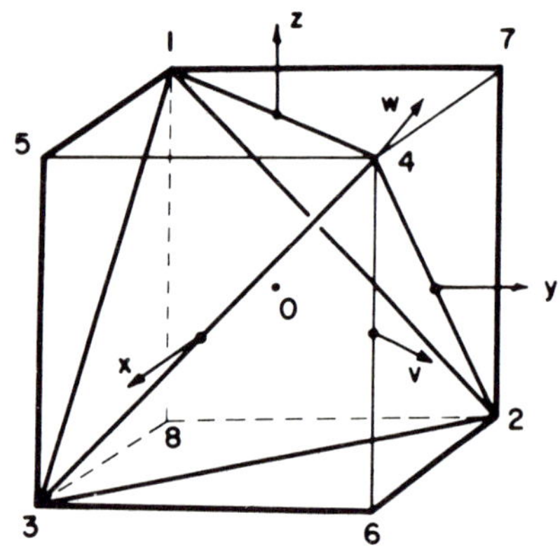

Figure 4. An illustration for the groups T, T_h, T_d, O, and O_h.

4.1. Character Tables and Basis Functions

For each group the first table given is the character table for the irreducible representations of the single and double group. Often the same character table serves for more than one group, because of an isomorphism. Only one group of an isomorphic set will contain only pure rotations. The other groups in the set may be obtained from this group by the replacement of selected pure rotations by the inversion times these rotations. In the character tables the horizontal line divides the single group representations $[\chi(\bar{R}) = \chi(R)]$ from the double group representations $[\chi(\bar{R}) = -\chi(R)]$. For even numbers of electrons one must use the representations above the line; for odd numbers of electrons one uses the representations below the line. For each irreducible representation, we indicate how the complex conjugate representation is related to the original representation (whether it corresponds to case a, b, or c of the time reversal conventions). The bases for the irreducible representations we have used in the construction of the coupling coefficients are also given. By a basis function $\phi(J,m)$ we mean a function which transforms like an eigenstate of the angular momentum operator **J** of total angular momentum J and z component m. Under proper rotations these functions transform according to the matrices $D_J(\alpha\beta\gamma)$ as defined by Wigner.[11] Under the inversion these functions are taken to go into themselves. By S_x, S_y, S_z we mean a basis transforming like like x, y, and z, except that they do not change sign under the inversion. By the symbol R we mean a function going into itself under all proper and improper rotations. By the notation $\phi(J,m)\times\Gamma_1^-$ (for the group D_{2h} for example) we mean a function transforming like $\phi(J,m)$ multiplied by a function transforming like a one-dimensional representation Γ_1^-. By a basis Γ_α^* we mean that this representation has as a basis the time inverses of the functions forming a basis for the representation Γ_α. For all of the bases given, we have merely used examples that show the transformation properties under spatial operators. In every case, no matter what basis is shown, the basis is assumed to have the time inversion properties listed in Section 3.3.

4.2. Multiplication Tables

Following the character table for each group, we give the representation multiplication table for the irreducible representations of that group. This table gives the coefficients $c_{\alpha\beta\gamma}$ (Equation 3.7) that list the number of times the γ irreducible representation is contained in the decomposition of the direct product of the α and β irreducible representations. For example, for the group O, $\Gamma_4 \times \Gamma_8 = \Gamma_6 + \Gamma_7 + 2\Gamma_8$ meaning that Γ_6 and Γ_7 are contained once and Γ_8 twice in the decomposition of the direct product of Γ_4 and Γ_8 ($c_{486} = 1$, $c_{487} = 1$, $c_{488} = 2$). The multiplication tables for groups constructed by supplementing a group already discussed with the inversion have been omitted. For these groups the bases must transform either symmetrically or antisymmetrically under inversion. In the first case, the representations are called + and in the second −. For these groups the $c_{\alpha\beta\gamma}$'s are the same as for the group before supplementing with the inversion, except that

$$\left.\begin{aligned} \Gamma_\alpha^\pm \times \Gamma_\beta^\pm &= \Sigma_\gamma c_{\alpha\beta\gamma}\Gamma_\gamma^+ \\ \Gamma_\alpha^\pm \times \Gamma_\beta^\mp &= \Sigma_\gamma c_{\alpha\beta\gamma}\Gamma_\gamma^- \end{aligned}\right\} \qquad (4.1)$$

4.3. Coupling Coefficients

In the tables following the multiplication table, we list all but the trivial coupling coefficients for the group in question. In the table of coupling coefficients, the columns are labeled by the product functions and the rows by the irreducible linear combinations. In the case of one-dimensional representations, the subscript denotes the irreducible representation (for example, u_1 for the cubic group). For multidimensional representations, the number of the irreducible representation is the superscript and the various partners are denoted by subscripts. A subscript +1 or −1, in this case, means a function transforming like $-i(x + iy)$ and $i(x - iy)$, or $-(S_x + iS_y)$ and $(S_x - iS_y)$. As the coefficients $U_{ijn}^{\alpha\beta\gamma}$ appear on the table, ij are the indices that run across a row, and the pair γn run down a column.

A function transforming according to the nth row of the irreducible representation can occur as a row label more than once for two reasons: because of time inversion (see the group C_1) or because $c_{\alpha\beta\gamma} > 1$. In the first case, we label all of the functions transforming in this way by ψ_n^γ. If $c_{\alpha\beta\gamma} = 2$, we call one linear combination ψ_n^γ and the other ϕ_n^γ. (See the group O as an example of this case.) For groups constructed by supplementing a group already discussed with the inversion, the coupling coefficients are also omitted. The coupling coefficients are just those given for the group before we supplement with the inversion. That is,

$$\left.\begin{aligned} \psi_n^{\gamma^+} &= \Sigma U_{ij,n}^{\alpha\beta,\gamma}\, u_i^{\alpha^\pm}\, V_j^{\beta^\pm} \\ \psi_n^{\gamma^-} &= \Sigma U_{ij,n}^{\alpha\beta,\gamma}\, u_i^{\alpha^\pm}\, V_j^{\beta^\pm} \end{aligned}\right\} \qquad (4.2)$$

Here the **U**'s are those for the group without the inversion.

4.4. Subgroup Compatibility Tables

The next table for each group concerns itself with the relation between the irreducible representations of a given group and those of the various subgroups of that group. In Figure 5, all of the thirty-two point groups are illustrated along with the possible subgroups of each of these groups. A heavy line in this figure indicates that the subgroup is not an invariant subgroup; a light line indicates that the subgroup is invariant. Consider a set of elements of the group that forms a subgroup. The set of matrices of a given irreducible representation of the group which represents these elements will form a representation of the subgroup. This representation may or may not be irreducible. If it is not, we may ask how it decomposes into irreducible representations of this subgroup. The number of times the αth irreducible representation of the subgroup is contained in the representation $\underset{\sim}{\Gamma}(R)$ of the group is given by

$$a_\alpha = \frac{1}{g_\alpha}\Sigma(R)\chi_\alpha^*(R)\chi(R) \qquad (4.3)$$

Here the sum extends over the elements of the subgroup the order of which is g_α. The numbers a_α are given in the compatibility tables. All the compatibility tables not tabulated are obtained by noting that if a group A is a subgroup of a group B, and B is a subgroup of the group C, then A is a subgroup of C.

The tables also contain the subgroups obtained by the reduction in symmetry caused by the application of an external constant electric or magnetic field to a system whose symmetry is initially described by one of the thirty-two point groups. The introduction of a constant external field will, in general, reduce the group of the Schrödinger equation for a given physical system to a subgroup of the original group. The symmetry of the external magnetic field is $C_{\infty h}$, while that of an external electric field is $C_{\infty v}$. The symmetry axis is the direction of the field. The group of the perturbed system consists of all

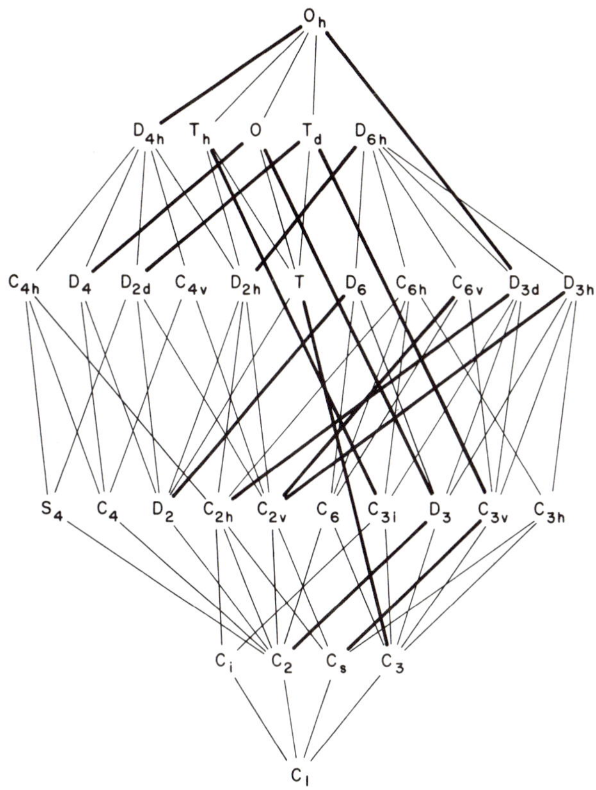

Figure 5. Subgroup decomposition of the thirty-two point groups. A heavy line indicates that the subgroup is not invariant.

operators that are members both of the original group and of the group of the external field. This will in general depend on the orientation of the external field with respect to the unperturbed group operations. In general with a field applied in an arbitrary direction the group is reduced to one of the trivial groups C_1, C_s, or C_i. Usually those field directions that so reduce the group are not included in the compatibility tables. The directions of the field are referred to in the figure illustrating the various groups.

1.5. Full Rotation Group Compatibility Tables

The final table for each group is similar to the compatibility tables which we described earlier. These arise from the observation that all of the thirty-two crystallographic point groups are subgroups of the full spherical group including all proper and improper rotations in three dimensions. The irreducible representations of this group are well known. They can be taken in the form given by Wigner[11] and are the matrices $\mathbf{D}_J^{\pm}(\alpha\beta\gamma)$; the J denotes the irreducible representation (total angular momentum) and can take on integral and half-integral values. The Euler angles describing the proper rotation are α, β, and γ (improper rotations are considered as being the inversion times some proper rotation described by these Euler angles), and the ± sign indicates whether the representation is an even or odd representation under the inversion. The integral values of J give the single group representations, and the half-integral values of J give the double group representations,. Each of the representations of this full spherical group may be decomposed into irreducible representations of each of the thirty-two point groups. All we need know to determine how one of these representations breaks up is a character table (Table 2) for the full rotation group. (Equation 4.3 can again be applied to this case.) The characters of the representations of the full rotation group are given by the formula

$$\chi\,[\mathbf{D}_J^{\pm}(R_\theta)] = \frac{\sin\left(J + \frac{1}{2}\right)\theta}{\sin\left(\frac{\theta}{2}\right)} \qquad (4.4)$$

for the proper rotations. Here R_θ is a rotation through an angle θ about some axis (every proper rotation is a rotation about some axis through some angle). The character of a proper rotation times the inversion is the same as the character of the proper rotation for the + representations and the negative of it for the – representations. The full rotation group compatibility tables indicate how the irreducible representations $\mathbf{D}_J^{\pm}(\alpha\beta\gamma)$ break up into irreducible representations of the various subgroups.

Chapter 5

USE OF THE TABLES

The application of the information contained in all of the tables, except possibly the tables of coupling coefficients, should be obvious from their definitions. The use of the coupling coefficient tables for finding linear combinations of the product functions which transform irreducibly is also straightforward. Take the entry in a given row of the table, multiply it by the column heading, and sum over the row to get the normalized function ψ_n^γ.

To use the tables of coupling coefficients to find the matrix elements of operators, we proceed as follows: Imagine that in the column headings we replace u_i^α by the operators $\mathbf{P}_i^\alpha$. The complex conjugate of the block of the matrix $\mathbf{U}$ given by ψ_n^γ or ϕ_n^γ, $n = 1 \cdots n_\gamma$, and $\mathbf{P}_i^\alpha \cdot v_j^\beta$, $j = 1 \cdots n_\beta$, gives the relations between the matrix elements $(u_n^\gamma \mid \mathbf{P}_i^\alpha \mid v_j^\beta)$. As an example, suppose that we want the matrix of an operator transforming like $x(\Gamma_4)$ under the group O (call the operator $\mathbf{P}_x^4$) between functions transforming like Γ_8 under the same group. We take the entries in the first four rows and columns in the 4×8 table of coupling coefficients for O and those in the second four rows and first four columns as our $\mathbf{U}$ and $\mathbf{U}'$ of Equation 3.24 ($c_{488} = 2$) and take their complex conjugate. We obtain then for

$$(u_n^8 \mid \mathbf{P}_x^4 \mid v_j^8)$$

the matrix

$$c_1 \begin{bmatrix} 0 & \frac{-i}{\sqrt{5}} & 0 & 0 \\ \frac{i}{\sqrt{5}} & 0 & \frac{-i\sqrt{4}}{\sqrt{15}} & 0 \\ 0 & \frac{-i\sqrt{4}}{\sqrt{15}} & 0 & \frac{-i}{\sqrt{5}} \\ 0 & 0 & \frac{-i}{\sqrt{5}} & 0 \end{bmatrix}$$

$$+ c_2 \begin{bmatrix} 0 & \frac{-i}{\sqrt{20}} & 0 & \frac{i\sqrt{5}}{\sqrt{12}} \\ \frac{-i}{\sqrt{20}} & 0 & \frac{i\sqrt{3}}{\sqrt{20}} & 0 \\ 0 & \frac{i\sqrt{3}}{\sqrt{20}} & 0 & \frac{-i}{\sqrt{20}} \\ \frac{i\sqrt{5}}{\sqrt{12}} & 0 & \frac{-i}{\sqrt{20}} & 0 \end{bmatrix}$$

Notice that in this case if the operator P is Hermitian and the u's are the same as the v's the constants (c's) necessarily are pure imaginary. In this case, in order to describe all of the matrix elements of an operator transforming like x we need two independent constants in general. The operators $\mathbf{P}_y^4$ and $\mathbf{P}_z^4$, which are partners with $\mathbf{P}_x^4$ in the $\underset{\sim}{\Gamma}_4$ irreducible representation, have matrix elements expressible in terms of the same two constants and the appropriate blocks of our coupling coefficient tables. It is only in the case where the ψ's and ϕ's appear as row labels that one obtains two matrices to express the elements of operators as in the illustration just described. If ψ occurs more than once as a row label, it is immaterial which set of rows one uses. As an example, consider the group T. Suppose we have an operator $\mathbf{P}^2$ transforming like $\underset{\sim}{\Gamma}_2$ and want to calculate the matrix elements of this operator between functions transforming like $\underset{\sim}{\Gamma}_4$. We can use either the matrix

$$\begin{bmatrix} \frac{1}{\sqrt{2}} & 0 & 0 \\ 0 & \frac{\omega^2}{\sqrt{2}} & 0 \\ 0 & 0 & \frac{\omega}{\sqrt{2}} \end{bmatrix}$$

where $\omega = \exp(2\pi i/3)$ for P^2, or the matrix

$$\begin{bmatrix} \frac{-i}{\sqrt{2}} & 0 & 0 \\ 0 & \frac{-i\omega^2}{\sqrt{2}} & 0 \\ 0 & 0 & \frac{-i\omega}{\sqrt{2}} \end{bmatrix}$$

since one matrix is just a constant times the other.

We thus have in the tables in Part 2 a complete description of the properties of the thirty-two point groups such as are useful in the discussion of crystal eigenstates. From the character tables, multiplication tables, and compatibility tables we can obtain a classification of these eigenstates according to the irreducible representations of the symmetry group of the crystal Hamiltonian and the selection rules that govern transitions between these states. The basis functions give a description of the symmetry properties of the eigenstates, and the coupling coefficients allow us to obtain that part of the matrix elements between crystal eigenstates which depends only on the symmetry of the system involved. That is, we have available all the information derivable solely from symmetry arguments concerning the eigenstates of a Hamiltonian, which is invariant under one of the thirty-two point groups.

REFERENCES

1. Bethe, H., Ann. Physik, 3, 133 (1929).
2. Eyring, H., J. Walter, and G.E. Kimball, Quantum Chemistry, John Wiley & Sons, Inc., New York, 1944, p. 383.
3. Landau, L.D., and E.M. Lifshitz, Quantum Mechanics – Non Relativistic Theory, Addison-Wesley Publishing Co., Inc., Reading, Mass., 1958, p. 338.
4. Margenau, H., and G.M. Murphy, The Mathematics of Physics and Chemistry, D. Van Nostrand Co.. Inc., New York, 1956, p. 579.
5. Elliott, R.J., Phys. Rev., 96, 280 (1954).
6. Parmenter, R.H., Phys. Rev., 100, 573 (1955).
7. Dresselhaus, G., Phys. Rev., 100, 580 (1955).
8. Glasser, M.L., J. Phys. Chem. Solids, 10, 229 (1959).
9. Quelle, F.W., Jr., Quarterly Progress Report Solid State and Molecular Theory Group M.I.T., Cambridge, Mass. (July 15, 1958), (unpublished).
10. Koster, G.F., "Space Groups and their Representations" Solid State Physics, Vol. 5, F. Seitz and D. Turnbull, Eds., Academic Press, Inc., New York, 1957, pp. 173-256.
11. Wigner, E.P., Group Theory, Academic Press, Inc., New York, 1959, Chapter 20.
12. Opechowski, W., Physica, 7, 552 (1940).
13. Rose, M.E., Elementary Theory of Angular Momentum, John Wiley & Sons, Inc., New York, 1957.
14. See Reference 11 for the definition of the Euler angles that we are using here.
15. Wigner, E.P., Nachr. Akad. Wiss. Göttingen, II Math-Physik. Kl., 546 (1932).
16. This relation is equivalent to that used in Reference 15.
17. Koster, G.F., Phys. Rev., 109, 227 (1958); Koster, G.F., and H. Statz, Phys. Rev., 113, 445 (1959); Statz, H., and G.F. Koster, Phys. Rev., 115, 1568 (1959).
18. The authors are indebted to V. Heine for pointing out this derivation which is more straightforward than the one contained in Reference 17.
19. International Tables for X-Ray Crystallography, Vol. I, Symmetry Groups, N.F.M. Henry and K. Lonsdale, Eds., Kynoch Press, Birmingham, England, 1952.

Part 2

TABLES

Table 1. Designation of the Thirty-Two Point Groups*

Notation

Point Group	International		Schön-flies Symbol	Point Group	International		Schön-flies Symbol
	Short	Full			Short	Full	
1	1	1	C_1	17	$\bar{3}$	$\bar{3}$	$C_{3i}(S_6)$
2	$\bar{1}$	$\bar{1}$	$C_i(S_2)$	18	32	32	D_3
3	2	2	C_2	19	3 M	3 M	C_{3v}
4	M	M	$C_s(C_{1h})$	20	$\bar{3}$ M	$\bar{3}\,\frac{2}{M}$	D_{3d}
5	2/M	$\frac{2}{M}$	C_{2h}	21	6	6	C_6
6	222	222	$D_2(V)$	22	$\bar{6}$	$\bar{6}$	C_{3h}
7	M M 2	M M 2	C_{2v}	23	6/M	$\frac{6}{M}$	C_{6h}
8	M M M	$\frac{2}{M}\,\frac{2}{M}\,\frac{2}{M}$	$D_{2h}(V_h)$	24	622	622	D_6
9	4	4	C_4	25	6 M M	6 M M	C_{6v}
10	$\bar{4}$	$\bar{4}$	S_4	26	$\bar{6}$ M 2	$\bar{6}$ M 2	D_{3h}
11	4/M	$\frac{4}{M}$	C_{4h}	27	6/MMM	$\frac{6}{M}\,\frac{2}{M}\,\frac{2}{M}$	D_{6h}
12	422	422	D_4	28	23	23	T
13	4 M M	4 M M	C_{4v}	29	M 3	$\frac{2}{M}\,\bar{3}$	T_h
14	$\bar{4}$ 2 M	$\bar{4}$ 2 M	$D_{2d}(V_d)$	30	432	432	O
15	4/M M M	$\frac{4}{M}\,\frac{2}{M}\,\frac{2}{M}$	D_{4h}	31	$\bar{4}$ 3 M	$\bar{4}$ 3 M	T_d
16	3	3	C_3	32	M 3 M	$\frac{4}{M}\,\bar{3}\,\frac{2}{M}$	O_h

*(See Reference 20, page 44.)

Table 2. Character Table for the Point Group Operators in the Full Rotation Group Representations

	E	C_6	C_4	C_3	C_2	I	S_3	S_4	S_6	σ
$D_0^\pm$	1	1	1	1	1	±1	±1	±1	±1	±1
$D_1^\pm$	3	2	1	0	−1	±3	±2	±1	0	∓1
$D_2^\pm$	5	1	−1	−1	1	±5	±1	∓1	∓1	±1
$D_3^\pm$	7	−1	−1	1	−1	±7	∓1	∓1	±1	∓1
$D_4^\pm$	9	−2	1	0	1	±9	∓2	±1	0	±1
$D_5^\pm$	11	−1	1	−1	−1	±11	∓1	±1	∓1	∓1
$D_6^\pm$	13	1	−1	1	1	±13	±1	∓1	±1	±1
$D_{1/2}^\pm$	2	$\sqrt{3}$	$\sqrt{2}$	1	0	±2	$\pm\sqrt{3}$	$\pm\sqrt{2}$	±1	0
$D_{3/2}^\pm$	4	$\sqrt{3}$	0	−1	0	±4	$\pm\sqrt{3}$	0	∓1	0
$D_{5/2}^\pm$	6	0	$-\sqrt{2}$	0	0	±6	0	$\mp\sqrt{2}$	0	0
$D_{7/2}^\pm$	8	$-\sqrt{3}$	0	1	0	±8	$\mp\sqrt{3}$	0	±1	0
$D_{9/2}^\pm$	10	$-\sqrt{3}$	$\sqrt{2}$	−1	0	±10	$\mp\sqrt{3}$	$\pm\sqrt{2}$	∓1	0
$D_{11/2}^\pm$	12	0	0	0	0	±12	0	0	0	0
$D_{13/2}^\pm$	14	$\sqrt{3}$	$-\sqrt{2}$	1	0	±14	$\pm\sqrt{3}$	$\mp\sqrt{2}$	±1	0

(1) C_1. This group consists of the identity E alone.

Table 3. Character Table for the Group C_1

C_1	E	$\bar{E}$	Time Inv.
Γ_1	1	1	a
Γ_2	1	−1	a

Table 4. Multiplication Table for the Group C_1

Γ_1	Γ_2	
Γ_1	Γ_2	Γ_1
	Γ_1	Γ_2

Table 5. Coupling Coefficients for the Group C_1

	$u_2 v_2$	$u_2 K v_2$	$(K u_2) v_2$	$K u_2 (K v_2)$
ψ_1	$1/\sqrt{2}$	0	0	$1/\sqrt{2}$
ψ_1	$i/\sqrt{2}$	0	0	$-i/\sqrt{2}$
ψ_1	0	$1/\sqrt{2}$	$-1/\sqrt{2}$	0
ψ_1	0	$i/\sqrt{2}$	$i/\sqrt{2}$	0

(2) C_i. This group consists of the identity and the inversion I.

Table 6. Character Table and Basis Functions for the Group C_i

C_i	E	$\bar{E}$	I	$\bar{I}$	Time Inv.	Bases
Γ_1^+	1	1	1	1	a	S_x or S_y or S_z
Γ_1^-	1	1	−1	−1	a	x or y or z
Γ_2^+	1	−1	1	−1	a	$\phi(1/2,-1/2)$ or $\phi(1/2, 1/2)$
Γ_2^-	1	−1	−1	1	a	$\Gamma_2^+ \times \Gamma_1^-$

Table 7. Compatibility Table for the Group C_i

C_i	Γ_1^+	Γ_1^-	Γ_2^+	Γ_2^-
C_1	Γ_1	Γ_1	Γ_2	Γ_2

Table 8. Multiplication Table for the Group C_i

Γ_1^+	Γ_1^-	Γ_2^+	Γ_2^-	
Γ_1^+	Γ_1^-	Γ_2^+	Γ_2^-	Γ_1^+
	Γ_1^+	Γ_2^-	Γ_2^+	Γ_1^-
		Γ_1^+	Γ_1^-	Γ_2^+
			Γ_1^+	Γ_2^-

(3) C_2. This group consists of the identity and a twofold rotation (taken about the z axis).

(4) C_s. This group, isomorphic to C_2, consists of the identity and a reflection (taken in the xy plane).

Table 9. Character Table and Basis Functions for the Groups C_2 and C_s

C_2	E	$\bar{E}$	C_2	$\bar{C}_2$	Time Inv.	Bases for C_2	Bases for C_s
C_s	E	$\bar{E}$	σ	$\bar{\sigma}$			
Γ_1	1	1	1	1	a	S_z or z	S_z or x or y
Γ_2	1	1	−1	−1	a	S_x or S_y or x or y	S_x or S_y or z
Γ_3	1	−1	i	−i	b	$\phi(1/2, 1/2)$	$\phi(1/2, 1/2)$
Γ_4	1	−1	−i	i	b	$\phi(1/2, -1/2)$	$\phi(1/2, -1/2)$

Table 10. Multiplication Table for the Groups C_2 and C_s

Γ_1	Γ_2	Γ_3	Γ_4	
Γ_1	Γ_2	Γ_3	Γ_4	Γ_1
	Γ_1	Γ_4	Γ_3	Γ_2
		Γ_2	Γ_1	Γ_3
			Γ_2	Γ_4

Table 11. Coupling Coefficients for the Groups C_s and C_2

$$u_4 = \mathbf{K} u_3 ; \; u_3 = -\mathbf{K} u_4$$

	$u_3 v_3$	$u_4 v_4$
ψ_2	$1/\sqrt{2}$	$1/\sqrt{2}$
ψ_2	$i/\sqrt{2}$	$-i/\sqrt{2}$

	$u_3 v_4$	$u_4 v_3$
ψ_1	$1/\sqrt{2}$	$-1/\sqrt{2}$
ψ_1	$i/\sqrt{2}$	$i/\sqrt{2}$

Table 12. Compatibility Table

C_2	C_s	Γ_1	Γ_2	Γ_3	Γ_4
C_1	C_1	Γ_1	Γ_1	Γ_2	Γ_2

Table 13. Full Rotation Group Compatibility Table for the Group C_2

$D_0^{\pm}$	Γ_1
$D_1^{\pm}$	$\Gamma_1 + 2\Gamma_2$
$D_2^{\pm}$	$3\Gamma_1 + 2\Gamma_2$
$D_3^{\pm}$	$3\Gamma_1 + 4\Gamma_2$
$D_4^{\pm}$	$5\Gamma_1 + 4\Gamma_2$
$D_5^{\pm}$	$5\Gamma_1 + 6\Gamma_2$
$D_6^{\pm}$	$7\Gamma_1 + 6\Gamma_2$
$D_j^{\pm}$	$(j + \frac{1}{2})(\Gamma_3 + \Gamma_4)$

Table 14. Full Rotation Group Compatibility Table for the Group C_s

D_0^+	Γ_1	D_0^-	Γ_2
D_1^+	$\Gamma_1 + 2\Gamma_2$	D_1^-	$2\Gamma_1 + \Gamma_2$
D_2^+	$3\Gamma_1 + 2\Gamma_2$	D_2^-	$2\Gamma_1 + 3\Gamma_2$
D_3^+	$3\Gamma_1 + 4\Gamma_2$	D_3^-	$4\Gamma_1 + 3\Gamma_2$
D_4^+	$5\Gamma_1 + 4\Gamma_2$	D^-	$4\Gamma_1 + 5\Gamma_2$
D_5^+	$5\Gamma_1 + 6\Gamma_2$	D_5^-	$6\Gamma_1 + 5\Gamma_2$
D_6^+	$7\Gamma_1 + 6\Gamma_2$	D_6^-	$6\Gamma_1 + 7\Gamma_2$
$D_j^{\pm}$	$(j + \frac{1}{2})(\Gamma_3 + \Gamma_4)$		

(5) C_{2h}. This group consists of the identity, a twofold rotation (taken about z), a reflection in a plane perpendicular to this axis (xy plane), and the inversion. It can be considered as the direct product of the two groups C_2 and C_i ($C_{2h} = C_2 \times C_i$). It can also be considered as the direct product of C_2 and C_s.

Table 15. Character Table and Basis Functions for the Group C_{2h}

C_{2h}	E	$\bar{E}$	C_2	$\bar{C}_2$	I	$\bar{I}$	σ_h	$\bar{\sigma}_h$	Time Inv.	Bases
Γ_1^+	1	1	1	1	1	1	1	1	a	S_z
Γ_2^+	1	1	−1	−1	1	1	−1	−1	a	S_x or S_y
Γ_1^-	1	1	1	1	−1	−1	−1	−1	a	z
Γ_2^-	1	1	−1	−1	−1	−1	1	1	a	x or y
Γ_3^+	1	−1	i	−i	1	−1	i	−i	b	$\phi(1/2, 1/2)$
Γ_4^+	1	−1	−i	i	1	−1	−i	i	b	$\phi(1/2,-1/2)$
Γ_3^-	1	−1	i	−i	−1	1	−i	i	b	$\Gamma_3^+ \times \Gamma_1^-$
Γ_4^-	1	−1	−i	i	−1	1	i	−i	b	$\Gamma_4^+ \times \Gamma_1^-$

Table 16. Compatibility Table

C_{2h} : H(z)	Γ_1^+	Γ_2^+	Γ_1^-	Γ_2^-	Γ_3^+	Γ_4^+	Γ_3^-	Γ_4^-
C_i : H(z)	Γ_1^+	Γ_1^+	Γ_1^-	Γ_1^-	Γ_2^+	Γ_2^+	Γ_2^-	Γ_2^-
C_2 : E(z)	Γ_1	Γ_2	Γ_1	Γ_2	Γ_3	Γ_4	Γ_3	Γ_4
C_s : E(x)	Γ_1	Γ_2	Γ_2	Γ_1	Γ_3	Γ_4	Γ_4	Γ_3

(6) D_2. This group consists of the identity and three twofold rotations about mutually perpendicular axes which may be designated z, y, and x.

(7) C_{2v}. This group, isomorphic to the group D_2, consists of the identity, two reflections, σ_v and σ_v' through planes perpendicular to each other (taken as the planes xz and yz) and a twofold rotation about the line of intersection of the planes (z axis).

Table 17. Character Table and Basis Functions for the Groups D_2 and C_{2v}

D_2	E	$\bar{E}$	C_2, $\bar{C}_2$	C_2', $\bar{C}_2'$	C_2'', $\bar{C}_2''$			
C_{2v}	E	$\bar{E}$	C_2, $\bar{C}_2$	σ_v, $\bar{\sigma}_v$	σ_v', $\bar{\sigma}_v'$	Time Inv.	Bases for D_2	Bases for C_{2v}
Γ_1	1	1	1	1	1	a	R or xyz	z
Γ_2	1	1	−1	1	−1	a	S_y or y	S_y or x
Γ_3	1	1	1	−1	−1	a	S_z or z	S_z or xy
Γ_4	1	1	−1	−1	1	a	S_x or x	S_x or y
Γ_5	2	−2	0	0	0	c	$\phi(1/2, -1/2)$, $\phi(1/2, 1/2)$	$\phi(1/2, -1/2)$ $\phi(1/2, 1/2)$

Table 18. Multiplication Table for the Groups D_2 and C_{2v}

Γ_1	Γ_2	Γ_3	Γ_4	Γ_5	
Γ_1	Γ_2	Γ_3	Γ_4	Γ_5	Γ_1
	Γ_1	Γ_4	Γ_3	Γ_5	Γ_2
		Γ_1	Γ_2	Γ_5	Γ_3
			Γ_1	Γ_5	Γ_4
				$\Gamma_1 + \Gamma_2 + \Gamma_3 + \Gamma_4$	Γ_5

Table 19. Coupling Coefficients for the Groups D_2 and C_{2v}

	$u_2 v^5_{-1/2}$	$u_2 v^5_{1/2}$
$\psi^5_{-1/2}$	0	1
$\psi^5_{1/2}$	-1	0

	$u_3 v^5_{-1/2}$	$u_3 v^5_{1/2}$
$\psi^5_{-1/2}$	i	0
$\psi^5_{1/2}$	0	$-i$

	$u_4 v^5_{-1/2}$	$u_4 v^5_{1/2}$
$\psi^5_{-1/2}$	0	i
$\psi^5_{1/2}$	i	0

	$u^5_{-1/2} v^5_{-1/2}$	$u^5_{-1/2} v^5_{1/2}$	$u^5_{1/2} v^5_{-1/2}$	$u^5_{1/2} v^5_{1/2}$
ψ_1	0	$-1/\sqrt{2}$	$1/\sqrt{2}$	0
ψ_3	0	$-i/\sqrt{2}$	$-i/\sqrt{2}$	0
ψ_2	$1/\sqrt{2}$	0	0	$1/\sqrt{2}$
ψ_4	$-i/\sqrt{2}$	0	0	$i/\sqrt{2}$

Table 20. Compatibility Table

D_2	C_{2v}: E(z)	Γ_1	Γ_2	Γ_3	Γ_4	Γ_5
C_2 : H(z) : E(z)	C_2 : H(z)	Γ_1	Γ_2	Γ_1	Γ_2	$\Gamma_3 + \Gamma_4$
C_2 : H(x) : E(x)	C_s : H(x) : E(y)	Γ_1	Γ_2	Γ_2	Γ_1	$\Gamma_3 + \Gamma_4$
C_2 : H(y) : E(y)	C_s : H(y) : E(x)	Γ_1	Γ_1	Γ_2	Γ_2	$\Gamma_3 + \Gamma_4$

Table 21. Full Rotation Group Compatibility Table for the Group D_2

$D_0^{\pm}$	Γ_1
$D_1^{\pm}$	$\Gamma_2 + \Gamma_3 + \Gamma_4$
$D_2^{\pm}$	$2\Gamma_1 + \Gamma_2 + \Gamma_3 + \Gamma_4$
$D_3^{\pm}$	$\Gamma_1 + 2\Gamma_2 + 2\Gamma_3 + 2\Gamma_4$
$D_4^{\pm}$	$3\Gamma_1 + 2\Gamma_2 + 2\Gamma_3 + 2\Gamma_4$
$D_5^{\pm}$	$2\Gamma_1 + 3\Gamma_2 + 3\Gamma_3 + 3\Gamma_4$
$D_6^{\pm}$	$4\Gamma_1 + 3\Gamma_2 + 3\Gamma_3 + 3\Gamma_4$
$D_j^{\pm}$	$(j + \frac{1}{2})\ \Gamma_5$

Table 22. Full Rotation Group Compatibility Table for the Group C_{2v}

D_0^{+}	Γ_1	D_0^{-}	Γ_3
D_1^{+}	$\Gamma_2 + \Gamma_3 + \Gamma_4$	D_1^{-}	$\Gamma_1 + \Gamma_2 + \Gamma_4$
D_2^{+}	$2\Gamma_1 + \Gamma_2 + \Gamma_3 + \Gamma_4$	D_2^{-}	$\Gamma_1 + \Gamma_2 + 2\Gamma_3 + \Gamma_4$
D_3^{+}	$\Gamma_1 + 2\Gamma_2 + 2\Gamma_3 + 2\Gamma_4$	D_3^{-}	$2\Gamma_1 + 2\Gamma_2 + \Gamma_3 + 2\Gamma_4$
D_4^{+}	$3\Gamma_1 + 2\Gamma_2 + 2\Gamma_3 + 2\Gamma_4$	D_4^{-}	$2\Gamma_1 + 2\Gamma_2 + 3\Gamma_3 + 2\Gamma_4$
D_5^{+}	$2\Gamma_1 + 3\Gamma_2 + 3\Gamma_3 + 3\Gamma_4$	D_5^{-}	$3\Gamma_1 + 3\Gamma_2 + 2\Gamma_3 + 3\Gamma_4$
D_6^{+}	$4\Gamma_1 + 3\Gamma_2 + 3\Gamma_3 + 3\Gamma_4$	D_6^{-}	$3\Gamma_1 + 3\Gamma_2 + 4\Gamma_3 + 3\Gamma_4$
D_j^{+}	$(j + \frac{1}{2})\ \Gamma_5$	D_j^{-}	$(j + \frac{1}{2})\ \Gamma_5$

(8) D_{2h}. This group is the direct product of the groups D_2 and C_i ($D_{2h} = D_2 \times C_i$). It can also be considered as the direct product of D_2 and C_s; of C_{2v} and C_i or of C_{2v} and C_s. In addition to the operations of D_2 and C_i it contains reflections through the planes perpendicular to the three twofold rotations of D_2. These planes are mutually perpendicular and each contains two rotation axes (planes xy, xz, and yz).

Table 23. Character Table and Basis Functions for the Group D_{2h}

D_{2h}	E	$\bar{E}$	C_2 $\bar{C}_2$	C_2' $\bar{C}_2'$	C_2'' $\bar{C}_2''$	I	$\bar{I}$	σ_v $\bar{\sigma}_v$	σ_v' $\bar{\sigma}_v'$	σ_v'' $\bar{\sigma}_v''$	Time Inv.	Bases
Γ_1^+	1	1	1	1	1	1	1	1	1	1	a	R
Γ_2^+	1	1	−1	1	−1	1	1	−1	1	−1	a	S_y
Γ_3^+	1	1	1	−1	−1	1	1	1	−1	−1	a	S_z
Γ_4^+	1	1	−1	−1	1	1	1	−1	−1	1	a	S_x
Γ_1^-	1	1	1	1	1	−1	−1	−1	−1	−1	a	xyz
Γ_2^-	1	1	−1	1	−1	−1	−1	1	−1	1	a	y
Γ_3^-	1	1	1	−1	−1	−1	−1	−1	1	1	a	z
Γ_4^-	1	1	−1	−1	1	−1	−1	1	1	−1	a	x
Γ_5^+	2	−2	0	0	0	2	−2	0	0	0	c	$\phi(1/2, -1/2)$, $\phi(1/2, 1/2)$
Γ_5^-	2	−2	0	0	0	−2	2	0	0	0	c	$\Gamma_5^+ \times \Gamma_1^-$

Table 24. Compatibility Table

D_{2h}	Γ_1^+	Γ_2^+	Γ_3^+	Γ_4^+	Γ_1^-	Γ_2^-	Γ_3^-	Γ_4^-	Γ_5^+	Γ_5^-
C_{2h}: H(z)	Γ_1^+	Γ_2^+	Γ_1^+	Γ_2^+	Γ_1^-	Γ_2^-	Γ_1^-	Γ_2^-	$\Gamma_3^+ + \Gamma_4^+$	$\Gamma_3^- + \Gamma_4^-$
C_{2h}: H(x)	Γ_1^+	Γ_2^+	Γ_2^+	Γ_1^+	Γ_1^-	Γ_2^-	Γ_2^-	Γ_1^-	$\Gamma_3^+ + \Gamma_4^+$	$\Gamma_3^- + \Gamma_4^-$
C_{2h}: H(y)	Γ_1^+	Γ_1^+	Γ_2^+	Γ_2^+	Γ_1^-	Γ_1^-	Γ_2^-	Γ_2^-	$\Gamma_3^+ + \Gamma_4^+$	$\Gamma_3^- + \Gamma_4^-$
C_{2v}: E(z)	Γ_1	Γ_2	Γ_3	Γ_4	Γ_3	Γ_4	Γ_1	Γ_2	Γ_5	Γ_5
C_{2v}: E(x)	Γ_1	Γ_4	Γ_2	Γ_3	Γ_3	Γ_2	Γ_4	Γ_1	Γ_5	Γ_5
C_{2v}: E(y)	Γ_1	Γ_3	Γ_2	Γ_4	Γ_3	Γ_1	Γ_4	Γ_2	Γ_5	Γ_5
D_2	Γ_1	Γ_2	Γ_3	Γ_4	Γ_1	Γ_2	Γ_3	Γ_4	Γ_5	Γ_5

(9) C_4. This is a cyclic group consisting of the identity; a rotation through 90°, C_4; through −90°, C_4^{-1}; and through 180°, C_2 all about a given axis (taken as z).

(10) S_4. This is a cyclic group, isomorphic to C_4, consisting of the identity; the rotation reflection $S_4 = IC_4^{-1}$; a twofold rotation, C_2; and the operator $S_4^{-1} = IC_4$. All rotations are taken about the z axis.

Table 25. Character Table and Basis Functions for the Groups C_4 and S_4

C_4	E	$\bar{E}$	C_4	$\bar{C}_4$	C_2	$\bar{C}_2$	C_4^{-1}	$\bar{C}_4^{-1}$	Time Inv.	Bases for C_4	Bases for S_4
S_4	E	$\bar{E}$	S_4^{-1}	$\bar{S}_4^{-1}$	C_2	$\bar{C}_2$	S_4	$\bar{S}_4$			
Γ_1	1	1	1	1	1	1	1	1	a	z or S_z	S_z
Γ_2	1	1	−1	−1	1	1	−1	−1	a	xy	z or xy
Γ_3	1	1	i	i	−1	−1	−i	−i	b	$-i(x+iy)$ or $-(S_x+iS_y)$	$-(S_x+iS_y)$ or $i(x-iy)$
Γ_4	1	1	−i	−i	−1	−1	i	i	b	$i(x-iy)$ or (S_x-iS_y)	(S_x-iS_y) or $-i(x+iy)$
Γ_5	1	−1	ω	$-\omega$	i	−i	$-\omega^3$	ω	b	$\phi(1/2, 1/2)$	$\phi(1/2, 1/2)$
Γ_6	1	−1	$-\omega^3$	ω^3	−i	i	ω	$-\omega^3$	b	$\phi(1/2,-1/2)$	$\phi(1/2,-1/2)$
Γ_7	1	−1	$-\omega$	ω	i	−i	ω^3	$-\omega$	b	$\phi(3/2,-3/2)$	$\phi(3/2,-3/2)$
Γ_8	1	−1	ω^3	$-\omega^3$	−i	i	$-\omega$	ω^3	b	$\phi(3/2, 3/2)$	$\phi(3/2, 3/2)$

$\omega = \exp(\pi i/4)$

Table 26. Multiplication Table for the Groups C_4 and S_4

Γ_1	Γ_2	Γ_3	Γ_4	Γ_5	Γ_6	Γ_7	Γ_8	
Γ_1	Γ_2	Γ_3	Γ_4	Γ_5	Γ_6	Γ_7	Γ_8	Γ_1
	Γ_1	Γ_4	Γ_3	Γ_7	Γ_8	Γ_5	Γ_6	Γ_2
		Γ_2	Γ_1	Γ_8	Γ_5	Γ_6	Γ_7	Γ_3
			Γ_2	Γ_6	Γ_7	Γ_8	Γ_5	Γ_4
				Γ_3	Γ_1	Γ_4	Γ_2	Γ_5
					Γ_4	Γ_2	Γ_3	Γ_6
						Γ_3	Γ_1	Γ_7
							Γ_4	Γ_8

Table 27. Coupling Coefficients for the Groups C_4 and S_4

$$u_4 = \mathbf{K}u_3;\ u_3 = \mathbf{K}u_4$$

$$u_6 = \mathbf{K}u_5;\ u_5 = -\mathbf{K}u_6$$

$$u_7 = \mathbf{K}u_8;\ u_8 = -\mathbf{K}u_7$$

$u_2v_5 = -\psi_7$	$u_3v_5 = \psi_8$	$u_4v_5 = -\psi_6$
$u_2v_6 = \psi_8$	$u_3v_6 = \psi_5$	$u_4v_6 = \psi_7$
$u_2v_7 = \psi_5$	$u_3v_7 = \psi_6$	$u_4v_7 = \psi_8$
$u_2v_8 = -\psi_6$	$u_3v_8 = -\psi_7$	$u_4v_8 = \psi_5$
$u_5v_5 = \psi_3$	$u_6v_6 = \psi_4$	$u_7v_7 = \psi_3$
$u_5v_7 = \psi_4$	$u_6v_8 = -\psi_3$	$u_8v_8 = \psi_4$

	u_3v_3	u_4v_4
ψ_2	$1/\sqrt{2}$	$1/\sqrt{2}$
ψ_2	$i/\sqrt{2}$	$-i/\sqrt{2}$

	u_3v_4	u_4v_3
ψ_1	$1/\sqrt{2}$	$1/\sqrt{2}$
ψ_1	$i/\sqrt{2}$	$-i/\sqrt{2}$

	u_5v_6	u_6v_5
ψ_1	$1/\sqrt{2}$	$-1/\sqrt{2}$
ψ_1	$i/\sqrt{2}$	$i/\sqrt{2}$

	u_5v_8	u_6v_7
ψ_2	$1/\sqrt{2}$	$1/\sqrt{2}$
ψ_2	$i/\sqrt{2}$	$-i/\sqrt{2}$

	u_7v_8	u_8v_7
ψ_1	$1/\sqrt{2}$	$-1/\sqrt{2}$
ψ_1	$i/\sqrt{2}$	$i/\sqrt{2}$

Table 28. Compatibility Table

C_4 : H (z) : E(z)	S_4 : H(z)	Γ_1	Γ_2	Γ_3	Γ_4	Γ_5	Γ_6	Γ_7	Γ_8
C_2	C_2 : E(z)	Γ_1	Γ_1	Γ_2	Γ_2	Γ_3	Γ_4	Γ_3	Γ_4

Table 29. Full Rotation Group Compatibility Table for the Group C_4

$D_0^{\pm}$	Γ_1
$D_1^{\pm}$	$\Gamma_1 + \Gamma_3 + \Gamma_4$
$D_2^{\pm}$	$\Gamma_1 + 2\Gamma_2 + \Gamma_3 + \Gamma_4$
$D_3^{\pm}$	$\Gamma_1 + 2\Gamma_2 + 2\Gamma_3 + 2\Gamma_4$
$D_4^{\pm}$	$3\Gamma_1 + 2\Gamma_2 + 2\Gamma_3 + 2\Gamma_4$
$D_5^{\pm}$	$3\Gamma_1 + 2\Gamma_2 + 3\Gamma_3 + 3\Gamma_4$
$D_6^{\pm}$	$3\Gamma_1 + 4\Gamma_2 + 3\Gamma_3 + 3\Gamma_4$
$D_{1/2}^{\pm}$	$\Gamma_5 + \Gamma_6$
$D_{3/2}^{\pm}$	$\Gamma_5 + \Gamma_6 + \Gamma_7 + \Gamma_8$
$D_{5/2}^{\pm}$	$\Gamma_5 + \Gamma_6 + 2\Gamma_7 + 2\Gamma_8$
$D_{7/2}^{\pm}$	$2\Gamma_5 + 2\Gamma_6 + 2\Gamma_7 + 2\Gamma_8$
$D_{9/2}^{\pm}$	$3\Gamma_5 + 3\Gamma_6 + 2\Gamma_7 + 2\Gamma_8$
$D_{11/2}^{\pm}$	$3\Gamma_5 + 3\Gamma_6 + 3\Gamma_7 + 3\Gamma_8$
$D_{13/2}^{\pm}$	$3\Gamma_5 + 3\Gamma_6 + 4\Gamma_7 + 4\Gamma_8$

Table 30. Full Rotation Group Compatibility Table for the Group S_4

D_0^{+}	Γ_1	D_0^{-}	Γ_2
D_1^{+}	$\Gamma_1 + \Gamma_3 + \Gamma_4$	D_1^{-}	$\Gamma_2 + \Gamma_3 + \Gamma_4$
D_2^{+}	$\Gamma_1 + 2\Gamma_2 + \Gamma_3 + \Gamma_4$	D_2^{-}	$2\Gamma_1 + \Gamma_2 + \Gamma_3 + \Gamma_4$
D_3^{+}	$\Gamma_1 + 2\Gamma_2 + 2\Gamma_3 + 2\Gamma_4$	D_3^{-}	$2\Gamma_1 + \Gamma_2 + 2\Gamma_3 + 2\Gamma_4$
D_4^{+}	$3\Gamma_1 + 2\Gamma_2 + 2\Gamma_3 + 2\Gamma_4$	D_4^{-}	$2\Gamma_1 + 3\Gamma_2 + 2\Gamma_3 + 2\Gamma_4$
D_5^{+}	$3\Gamma_1 + 2\Gamma_2 + 3\Gamma_3 + 3\Gamma_4$	D_5^{-}	$2\Gamma_1 + 3\Gamma_2 + 3\Gamma_3 + 3\Gamma_4$
D_6^{+}	$3\Gamma_1 + 4\Gamma_2 + 3\Gamma_3 + 3\Gamma_4$	D_6^{-}	$4\Gamma_1 + 3\Gamma_2 + 3\Gamma_3 + 3\Gamma_4$
$D_{1/2}^{+}$	$\Gamma_5 + \Gamma_6$	$D_{1/2}^{-}$	$\Gamma_7 + \Gamma_8$
$D_{3/2}^{+}$	$\Gamma_5 + \Gamma_6 + \Gamma_7 + \Gamma_8$	$D_{3/2}^{-}$	$\Gamma_5 + \Gamma_6 + \Gamma_7 + \Gamma_8$
$D_{5/2}^{+}$	$\Gamma_5 + \Gamma_6 + 2\Gamma_7 + 2\Gamma_8$	$D_{5/2}^{-}$	$2\Gamma_5 + 2\Gamma_6 + \Gamma_7 + \Gamma_8$
$D_{7/2}^{+}$	$2\Gamma_5 + 2\Gamma_6 + 2\Gamma_7 + 2\Gamma_8$	$D_{7/2}^{-}$	$2\Gamma_5 + 2\Gamma_6 + 2\Gamma_7 + 2\Gamma_8$
$D_{9/2}^{+}$	$3\Gamma_5 + 3\Gamma_6 + 2\Gamma_7 + 2\Gamma_8$	$D_{9/2}^{-}$	$2\Gamma_5 + 2\Gamma_6 + 3\Gamma_7 + 3\Gamma_8$
$D_{11/12}^{+}$	$3\Gamma_5 + 3\Gamma_6 + 3\Gamma_7 + 3\Gamma_8$	$D_{11/2}^{-}$	$3\Gamma_5 + 3\Gamma_6 + 3\Gamma_7 + 3\Gamma_8$
$D_{13/2}^{+}$	$3\Gamma_5 + 3\Gamma_6 + 4\Gamma_7 + 4\Gamma_8$	$D_{13/2}^{-}$	$4\Gamma_5 + 4\Gamma_6 + 3\Gamma_7 + 3\Gamma_8$

(11) C_{4h}. This group is the direct product of the groups C_4 and C_i ($C_{4h} = C_4 \times C_i$). It can also be considered as the direct product of C_4 and C_s, of S_4 and C_i, or of S_4 and C_s. In addition to the operators of C_4 and C_1 it contains S_4, a reflection σ_h through a plane perpendicular to the axis of rotation (z axis, xy plane) and S_4^{-1}.

Table 31. Character Table and Basis Functions for the Group C_{4h}.

C_{4h}	E	$\bar{E}$	C_4	$\bar{C}_4$	C_2	$\bar{C}_2$	C_4^{-1}	$\bar{C}_4^{-1}$	I	$\bar{I}$	S_4^{-1}	$\bar{S}_4^{-1}$	σ_h	$\bar{\sigma}_h$	S_4	$\bar{S}_4$	Time Inv.	Bases
Γ_1^+	1	1	1	1	1	1	1	1	1	1	1	1	1	1	1	1	a	Sz
Γ_2^+	1	1	−1	−1	1	1	−1	−1	1	1	−1	−1	1	1	−1	−1	a	xy
Γ_3^+	1	1	i	i	−1	−1	−i	−i	1	1	i	i	−1	−1	−i	−i	b	$-(S_x + iS_y)$
Γ_4^+	1	1	−i	−i	−1	−1	i	i	1	1	−i	−i	−1	−1	i	i	b	$(S_x - iS_y)$
Γ_1^-	1	1	1	1	1	1	1	1	−1	−1	−1	−1	−1	−1	−1	−1	a	z
Γ_2^-	1	1	−1	−1	1	1	−1	−1	−1	−1	1	1	−1	−1	1	1	a	xyz
Γ_3^-	1	1	i	i	−1	−1	−i	−i	−1	−1	−i	−i	1	1	i	i	b	$-i(x+iy)$
Γ_4^-	1	1	−i	−i	−1	−1	i	i	−1	−1	i	i	1	1	−i	−i	b	$i(x-iy)$
Γ_5^+	1	−1	ω	$-\omega$	i	−i	$-\omega^3$	ω^3	1	−1	ω	$-\omega$	i	−i	$-\omega^3$	ω^3	b	$\phi(1/2, 1/2)$
Γ_6^+	1	−1	$-\omega^3$	ω^3	−i	i	ω	$-\omega$	1	−1	$-\omega^3$	ω^3	−i	−i	ω	$-\omega$	b	$\phi(1/2, -1/2)$
Γ_7^+	1	−1	$-\omega$	ω	i	−i	ω^3	$-\omega^3$	1	−1	$-\omega$	ω	i	−i	ω^3	$-\omega^3$	b	$\phi(3/2, -3/2)$
Γ_8^+	1	−1	ω^3	$-\omega^3$	−i	i	$-\omega$	ω	1	−1	ω^3	$-\omega^3$	−i	i	$-\omega$	ω	b	$\phi(3/2, 3/2)$
Γ_5^-	1	−1	ω	$-\omega$	i	−i	$-\omega^3$	ω^3	−1	1	$-\omega$	ω	−i	i	ω^3	$-\omega^3$	b	$\Gamma_5^+ \times \Gamma_1^-$
Γ_6^-	1	−1	$-\omega^3$	ω^3	−i	i	ω	$-\omega$	−1	1	ω^3	$-\omega^3$	i	−i	$-\omega$	ω	b	$\Gamma_6^+ \times \Gamma_1^-$
Γ_7^-	1	−1	$-\omega$	ω	i	−i	ω^3	$-\omega^3$	−1	1	ω	$-\omega$	−i	i	$-\omega^3$	ω^3	b	$\Gamma_7^+ \times \Gamma_1^-$
Γ_8^-	1	−1	ω^3	$-\omega^3$	−i	i	$-\omega$	ω	−1	1	$-\omega^3$	ω^3	i	−i	ω	$-\omega$	b	$\Gamma_8^+ \times \Gamma_1^-$

$\omega = \exp(\pi i/4)$

Table 32. Compatibility Table

C_{4h} : H(z)	Γ_1^+	Γ_2^+	Γ_3^+	Γ_4^+	Γ_1^-	Γ_2^-	Γ_3^-	Γ_4^-	Γ_5^+	Γ_6^+	Γ_7^+	Γ_8^+	Γ_5^-	Γ_6^-	Γ_7^-	Γ_8^-
C_4 : E(z)	Γ_1	Γ_2	Γ_3	Γ_4	Γ_1	Γ_2	Γ_3	Γ_4	Γ_5	Γ_6	Γ_7	Γ_8	Γ_5	Γ_6	Γ_7	Γ_8
S_4	Γ_1	Γ_2	Γ_3	Γ_4	Γ_2	Γ_1	Γ_4	Γ_3	Γ_5	Γ_6	Γ_7	Γ_8	Γ_7	Γ_8	Γ_5	Γ_6
C_{2h}	Γ_1^+	Γ_1^+	Γ_2^+	Γ_2^+	Γ_1^-	Γ_1^-	Γ_2^-	Γ_2^-	Γ_3^+	Γ_4^+	Γ_3^+	Γ_4^+	Γ_3^-	Γ_4^-	Γ_3^-	Γ_4^-
C_i : H(x):H(y)	Γ_1^+	Γ_1^+	Γ_1^+	Γ_1^+	Γ_1^-	Γ_1^-	Γ_1^-	Γ_1^-	Γ_2^+	Γ_2^+	Γ_2^+	Γ_2^+	Γ_2^-	Γ_2^-	Γ_2^-	Γ_2^-
C_s : E(x):E(y)	Γ_1	Γ_1	Γ_2	Γ_2	Γ_2	Γ_2	Γ_1	Γ_1	Γ_3	Γ_4	Γ_3	Γ_4	Γ_4	Γ_3	Γ_4	Γ_3

(12) D_4. This group consists of a fourfold axis z and four twofold axes perpendicular to the fourfold axis. The two twofold axes $C_{2'}$ are taken as the x and y axes; the two twofold axes $C_{2''}$ are taken as the axes 1 and 2 in Figure 2 which make 45° angles with the x and y axes and are mutually perpendicular.

(13) C_{4v}. This group, isomorphic to D_4, consists of a fourfold axis z and four reflection planes containing this axis. The two reflections labeled σ_v are taken in the planes xz and yz, while those labeled σ_d are taken in the planes z1 and z2.

(14) D_{2d}. This group, isomorphic to D_4, consists of the operations of D_2 and, in addition, of the operations S_4 and S_4^{-1} about one of the twofold axes z. There are also two reflections σ_d through perpendicular planes which contain the axis of S_4 and which bisect the angles between the two rotations of D_2 about the axes x and y, $C_{2'}$. The two reflection planes would be z1 and z2 in Figure 2.

Table 33. Character Table and Basis Functions for the Groups D_4, C_{4v}, and D_{2d}

D_4	E	$\bar{E}$	$2C_4$	$2\bar{C}_4$	C_2 $\bar{C}_2$	$2C_2'$ $2\bar{C}_2'$	$2C_2''$ $2\bar{C}_2''$				
C_{4v}	E	$\bar{E}$	$2C_4$	$2\bar{C}_4$	C_2 $\bar{C}_2$	$2\sigma_v$ $2\bar{\sigma}_v$	$2\sigma_d$ $2\bar{\sigma}_d$				
D_{2d}	E	$\bar{E}$	$2S_4$	$2\bar{S}_4$	C_2 $\bar{C}_2$	$2C_2'$ $2\bar{C}_2'$	$2\sigma_d$ $2\bar{\sigma}_d$	Time Inv.	Bases for D_4	Bases for C_{4v}	Bases for D_{2d}
Γ_1	1	1	1	1	1	1	1	a	R	R or z	R
Γ_2	1	1	1	1	1	−1	−1	a	z or S_z	S_z	S_z
Γ_3	1	1	−1	−1	1	1	−1	a	(x^2-y^2)	(x^2-y^2)	(x^2-y^2)
Γ_4	1	1	−1	−1	1	−1	1	a	xy	xy	xy or z
Γ_5	2	2	0	0	−2	0	0	a	S_x, S_y	S_x, S_y	S_x, S_y
Γ_6	2	−2	$\sqrt{2}$	$-\sqrt{2}$	0	0	0	c	$\phi(1/2,-1/2)$, $\phi(1/2,1/2)$	$\phi(1/2,-1/2)$, $\phi(1/2,1/2)$	$\phi(1/2,-1/2)$, $\phi(1/2,1/2)$
Γ_7	2	−2	$-\sqrt{2}$	$\sqrt{2}$	0	0	0	c	$\Gamma_6\times\Gamma_3$	$\Gamma_6\times\Gamma_3$	$\Gamma_6\times\Gamma_3$

Table 34. Multiplication Table for the Groups D_4, C_{4v}, and D_{2d}

Γ_1	Γ_2	Γ_3	Γ_4	Γ_5	Γ_6	Γ_7	
Γ_1	Γ_2	Γ_3	Γ_4	Γ_5	Γ_6	Γ_7	Γ_1
	Γ_1	Γ_4	Γ_3	Γ_5	Γ_6	Γ_7	Γ_2
		Γ_1	Γ_2	Γ_5	Γ_7	Γ_6	Γ_3
			Γ_1	Γ_5	Γ_7	Γ_6	Γ_4
				$\Gamma_1+\Gamma_2$ $+\Gamma_3+\Gamma_4$	$\Gamma_6+\Gamma_7$	$\Gamma_6+\Gamma_7$	Γ_5
					$\Gamma_1+\Gamma_2$ $+\Gamma_5$	$\Gamma_3+\Gamma_4$ $+\Gamma_5$	Γ_6
						$\Gamma_1+\Gamma_2$ $+\Gamma_5$	Γ_7

Table 35. Coupling Coefficients for the Groups D_4, C_{4v}, and D_{2d}

	$u_2v^5_x$	$u_2v^5_y$
ψ^5_x	0	1
ψ^5_y	−1	0

	$u_2v^6_{-1/2}$	$u_2v^6_{1/2}$
$\psi^6_{-1/2}$	i	0
$\psi^6_{1/2}$	0	−i

	$u_2v^7_{-1/2}$	$u_2v^7_{1/2}$
$\psi^7_{-1/2}$	i	0
$\psi^7_{1/2}$	0	−i

	$u_3v^5_x$	$u_3v^5_y$
ψ^5_x	0	1
ψ^5_y	1	0

	$u_3v^6_{-1/2}$	$u_3v^6_{1/2}$
$\psi^7_{-1/2}$	1	0
$\psi^7_{1/2}$	0	1

	$u_3v^7_{-1/2}$	$u_3v^7_{1/2}$
$\psi^6_{-1/2}$	1	0
$\psi^6_{1/2}$	0	1

	$u_4v^5_x$	$u_4v^5_y$
ψ^5_x	0	1
ψ^5_y	1	0

	$u_4v^6_{-1/2}$	$u_4v^6_{1/2}$
$\psi^7_{-1/2}$	i	0
$\psi^7_{1/2}$	0	−i

	$u_4v^7_{-1/2}$	$u_4v^7_{1/2}$
$\psi^6_{-1/2}$	i	0
$\psi^6_{1/2}$	0	−i

	$u^5_xv^5_x$	$u^5_xv^5_y$	$u^5_yv^5_x$	$u^5_yv^5_y$
ψ_1	$1/\sqrt{2}$	0	0	$1/\sqrt{2}$
ψ_2	0	$1/\sqrt{2}$	$-1/\sqrt{2}$	0
ψ_3	$1/\sqrt{2}$	0	0	$-1/\sqrt{2}$
ψ_4	0	$1/\sqrt{2}$	$1/\sqrt{2}$	0

	$u^5_xv^6_{-1/2}$	$u^5_xv^6_{1/2}$	$u^5_yv^6_{-1/2}$	$u^5_yv^6_{1/2}$
$\psi^6_{-1/2}$	0	$i/\sqrt{2}$	0	$-1/\sqrt{2}$
$\psi^6_{1/2}$	$i/\sqrt{2}$	0	$1/\sqrt{2}$	0
$\psi^7_{-1/2}$	0	$i/\sqrt{2}$	0	$1/\sqrt{2}$
$\psi^7_{1/2}$	$i/\sqrt{2}$	0	$-1/\sqrt{2}$	0

	$u^5_xv^7_{-1/2}$	$u^5_xv^7_{1/2}$	$u^5_yv^7_{-1/2}$	$u^5_yv^7_{1/2}$
$\psi^6_{-1/2}$	0	$i/\sqrt{2}$	0	$1/\sqrt{2}$
$\psi^6_{1/2}$	$i/\sqrt{2}$	0	$-1/\sqrt{2}$	0
$\psi^7_{-1/2}$	0	$i/\sqrt{2}$	0	$-1/\sqrt{2}$
$\psi^7_{1/2}$	$i/\sqrt{2}$	0	$1/\sqrt{2}$	0

Table 35. Coupling Coefficients for the Groups D_4, C_{4v}, and D_{2d} (Cont.)

	$u^6_{-1/2}v^6_{-1/2}$	$u^6_{-1/2}v^6_{1/2}$	$u^6_{1/2}v^6_{-1/2}$	$u^6_{1/2}v^6_{1/2}$
ψ_1	0	$-1/\sqrt{2}$	$1/\sqrt{2}$	0
ψ_2	0	$-i/\sqrt{2}$	$-i/\sqrt{2}$	0
ψ^5_x	$-i/\sqrt{2}$	0	0	$i/\sqrt{2}$
ψ^5_y	$1/\sqrt{2}$	0	0	$1/\sqrt{2}$

	$u^6_{-1/2}v^7_{-1/2}$	$u^6_{-1/2}v^7_{1/2}$	$u^6_{1/2}v^7_{-1/2}$	$u^6_{1/2}v^7_{1/2}$
ψ_3	0	$-1/\sqrt{2}$	$1/\sqrt{2}$	0
ψ_4	0	$-i/\sqrt{2}$	$-i/\sqrt{2}$	0
ψ^5_x	$1/\sqrt{2}$	0	0	$1/\sqrt{2}$
ψ^5_y	$-i/\sqrt{2}$	0	0	$i/\sqrt{2}$

	$u^7_{-1/2}v^7_{-1/2}$	$u^7_{-1/2}v^7_{1/2}$	$u^7_{1/2}v^7_{-1/2}$	$u^7_{1/2}v^7_{1/2}$
ψ_1	0	$-1/\sqrt{2}$	$1/\sqrt{2}$	0
ψ_2	0	$-i/\sqrt{2}$	$-i/\sqrt{2}$	0
ψ^5_x	$-i/\sqrt{2}$	0	0	$i/\sqrt{2}$
ψ^5_y	$1/\sqrt{2}$	0	0	$1/\sqrt{2}$

Table 36. Compatibility Table

D_4	C_{4v}: E(z)	D_{2d}	Γ_1	Γ_2	Γ_3	Γ_4	Γ_5	Γ_6	Γ_7
C_4: H(z):E(z)	C_4:H(z)	S_4:H(z)	Γ_1	Γ_1	Γ_2	Γ_2	$\Gamma_3+\Gamma_4$	$\Gamma_5+\Gamma_6$	$\Gamma_7+\Gamma_8$
C_2: H(x):E(x)	C_s: H(x):E(y)	C_2:H(x):E(y)	Γ_1	Γ_2	Γ_1	Γ_2	$\Gamma_1+\Gamma_2$	$\Gamma_3+\Gamma_4$	$\Gamma_3+\Gamma_4$
D_2	C_{2v}	D_2	Γ_1	Γ_3	Γ_1	Γ_3	$\Gamma_2+\Gamma_4$	Γ_5	Γ_5
		C_{2v} : E(z)	Γ_1	Γ_3	Γ_3	Γ_1	$\Gamma_2+\Gamma_4$	Γ_5	Γ_5

Table 37. Full Rotation Group Compatibility Table for the Group D_4

$D_0^\pm$	Γ_1
$D_1^\pm$	$\Gamma_2 + \Gamma_5$
$D_2^\pm$	$\Gamma_1 + \Gamma_3 + \Gamma_4 + \Gamma_5$
$D_3^\pm$	$\Gamma_2 + \Gamma_3 + \Gamma_4 + 2\Gamma_5$
$D_4^\pm$	$2\Gamma_1 + \Gamma_2 + \Gamma_3 + \Gamma_4 + 2\Gamma_5$
$D_5^\pm$	$\Gamma_1 + 2\Gamma_2 + \Gamma_3 + \Gamma_4 + 3\Gamma_5$
$D_6^\pm$	$2\Gamma_1 + \Gamma_2 + 2\Gamma_3 + 2\Gamma_4 + 3\Gamma_5$
$D_{1/2}^\pm$	Γ_6
$D_{3/2}^\pm$	$\Gamma_6 + \Gamma_7$
$D_{5/2}^\pm$	$\Gamma_6 + 2\Gamma_7$
$D_{7/2}^\pm$	$2\Gamma_6 + 2\Gamma_7$
$D_{9/2}^\pm$	$3\Gamma_6 + 2\Gamma_7$
$D_{11/2}^\pm$	$3\Gamma_6 + 3\Gamma_7$
$D_{13/2}^\pm$	$3\Gamma_6 + 4\Gamma_7$

Table 38. Full Rotation Group Compatibility Table for the Group C_{4v}

D_0^+	Γ_1	D_0^-	Γ_2
D_1^+	$\Gamma_2 + \Gamma_5$	D_1^-	$\Gamma_1 + \Gamma_5$
D_2^+	$\Gamma_1 + \Gamma_3 + \Gamma_4 + \Gamma_5$	D_2^-	$\Gamma_2 + \Gamma_3 + \Gamma_4 + \Gamma_5$
D_3^+	$\Gamma_2 + \Gamma_3 + \Gamma_4 + 2\Gamma_5$	D_3^-	$\Gamma_1 + \Gamma_3 + \Gamma_4 + 2\Gamma_5$
D_4^+	$2\Gamma_1 + \Gamma_2 + \Gamma_3 + \Gamma_4 + 2\Gamma_5$	D_4^-	$\Gamma_1 + 2\Gamma_2 + \Gamma_3 + \Gamma_4 + 2\Gamma_5$
D_5^+	$\Gamma_1 + 2\Gamma_2 + \Gamma_3 + \Gamma_4 + 3\Gamma_5$	D_5^-	$2\Gamma_1 + \Gamma_2 + \Gamma_3 + \Gamma_4 + 3\Gamma_5$
D_6^+	$2\Gamma_1 + \Gamma_2 + 2\Gamma_3 + 2\Gamma_4 + 3\Gamma_5$	D_6^-	$\Gamma_1 + 2\Gamma_2 + 2\Gamma_3 + 2\Gamma_4 + 3\Gamma_5$
$D_{1/2}^+$	Γ_6	$D_{1/2}^-$	Γ_6
$D_{3/2}^+$	$\Gamma_6 + \Gamma_7$	$D_{3/2}^-$	$\Gamma_6 + \Gamma_7$
$D_{5/2}^+$	$\Gamma_6 + 2\Gamma_7$	$D_{5/2}^-$	$\Gamma_6 + 2\Gamma_7$
$D_{7/2}^+$	$2\Gamma_6 + 2\Gamma_7$	$D_{7/2}^-$	$2\Gamma_6 + 2\Gamma_7$
$D_{9/2}^+$	$3\Gamma_6 + 2\Gamma_7$	$D_{9/2}^-$	$3\Gamma_6 + 2\Gamma_7$
$D_{11/2}^+$	$3\Gamma_6 + 3\Gamma_7$	$D_{11/2}^-$	$3\Gamma_6 + 3\Gamma_7$
$D_{13/2}^+$	$3\Gamma_6 + 4\Gamma_7$	$D_{13/2}^-$	$3\Gamma_6 + 4\Gamma_7$

Table 39. Full Rotation Group Compatibility Table for the Group D_{2d}

D_0^+	Γ_1	D_0^-	Γ_3
D_1^+	$\Gamma_2 + \Gamma_5$	D_1^-	$\Gamma_4 + \Gamma_5$
D_2^+	$\Gamma_1 + \Gamma_3 + \Gamma_4 + \Gamma_5$	D_2^-	$\Gamma_1 + \Gamma_2 + \Gamma_3 + \Gamma_5$
D_3^+	$\Gamma_2 + \Gamma_3 + \Gamma_4 + 2\Gamma_5$	D_3^-	$\Gamma_1 + \Gamma_2 + \Gamma_4 + 2\Gamma_5$
D_4^+	$2\Gamma_1 + \Gamma_2 + \Gamma_3 + \Gamma_4 + 2\Gamma_5$	D_4^-	$\Gamma_1 + \Gamma_2 + 2\Gamma_3 + \Gamma_4 + 2\Gamma_5$
D_5^+	$\Gamma_1 + 2\Gamma_2 + \Gamma_3 + \Gamma_4 + 3\Gamma_5$	D_5^-	$\Gamma_1 + \Gamma_2 + \Gamma_3 + 2\Gamma_4 + 3\Gamma_5$
D_6^+	$2\Gamma_1 + \Gamma_2 + 2\Gamma_3 + 2\Gamma_4 + 3\Gamma_5$	D_6^-	$2\Gamma_1 + 2\Gamma_2 + 2\Gamma_3 + \Gamma_4 + 3\Gamma_5$
$D_{1/2}^+$	Γ_6	$D_{1/2}^-$	Γ_7
$D_{3/2}^+$	$\Gamma_6 + \Gamma_7$	$D_{3/2}^-$	$\Gamma_6 + \Gamma_7$
$D_{5/2}^+$	$\Gamma_6 + 2\Gamma_7$	$D_{5/2}^-$	$2\Gamma_6 + \Gamma_7$
$D_{7/2}^+$	$2\Gamma_6 + 2\Gamma_7$	$D_{7/2}^-$	$2\Gamma_6 + 2\Gamma_7$
$D_{9/2}^+$	$3\Gamma_6 + 2\Gamma_7$	$D_{9/2}^-$	$2\Gamma_6 + 3\Gamma_7$
$D_{11/2}^+$	$3\Gamma_6 + 3\Gamma_7$	$D_{11/2}^-$	$3\Gamma_6 + 3\Gamma_7$
$D_{13/2}^+$	$3\Gamma_6 + 4\Gamma_7$	$D_{13/2}^-$	$4\Gamma_6 + 3\Gamma_7$

(15) D_{4h}. This group is the direct product of D_4 and C_i ($D_{4h} = D_4 \times C_i$). It can also be considered as the direct product of D_4 and C_s, of C_{4v} and C_i, of C_{4v} and C_s, of D_{2d} and C_i, or of D_{2d} and C_s. Referring to Figure 2, if we call z the fourfold axis of D_4, the operations other than those of D_4 and C_i are: S_4 and S_4^{-1}, the reflections σ_v through the xz and yz planes, the reflections σ_d through the planes z1 and z2, and the reflection σ_h through the plane xy perpendicular to the fourfold axis.

Table 40. Character Table and Basis Functions for the Group D_{4h}

D_{4h}	E	$\bar{E}$	$2C_4$	$2\bar{C}_4$	C_2 $\bar{C}_2$	$2C_2'$ $2\bar{C}_2'$	$2C_2''$ $2\bar{C}_2''$	I	$\bar{I}$	$2S_4$	$2\bar{S}_4$	σ_h $\bar{\sigma}_h$	$2\sigma_v$ $2\bar{\sigma}_v$	$2\sigma_d$ $2\bar{\sigma}_d$	Time Inv.	Bases
Γ_1^+	1	1	1	1	1	1	1	1	1	1	1	1	1	1	a	R
Γ_2^+	1	1	1	1	1	−1	−1	1	1	1	1	1	−1	−1	a	S_z
Γ_3^+	1	1	−1	−1	1	1	−1	1	1	−1	−1	1	1	−1	a	(x^2-y^2)
Γ_4^+	1	1	−1	−1	1	−1	1	1	1	−1	−1	1	−1	1	a	xy
Γ_5^+	2	2	0	0	−2	0	0	2	2	0	0	−2	0	0	a	S_x, S_y
Γ_1^-	1	1	1	1	1	1	1	−1	−1	−1	−1	−1	−1	−1	a	$(x^2-y^2)xyz$
Γ_2^-	1	1	1	1	1	−1	−1	−1	−1	−1	−1	−1	1	1	a	z
Γ_3^-	1	1	−1	−1	1	1	−1	−1	−1	1	1	−1	−1	1	a	xyz
Γ_4^-	1	1	−1	−1	1	−1	1	−1	−1	1	1	−1	1	−1	a	$(x^2-y^2)z$
Γ_5^-	2	2	0	0	−2	0	0	−2	−2	0	0	2	0	0	a	x,y
Γ_6^+	2	−2	$\sqrt{2}$	$-\sqrt{2}$	0	0	0	2	−2	$\sqrt{2}$	$-\sqrt{2}$	0	0	0	c	$\phi(1/2,-1/2)$, $\phi(1/2, 1/2)$
Γ_7^+	2	−2	$-\sqrt{2}$	$\sqrt{2}$	0	0	0	2	−2	$-\sqrt{2}$	$\sqrt{2}$	0	0	0	c	$\Gamma_6^+ \times \Gamma_3^+$
Γ_6^-	2	−2	$\sqrt{2}$	$-\sqrt{2}$	0	0	0	−2	2	$-\sqrt{2}$	$\sqrt{2}$	0	0	0	c	$\Gamma_6^+ \times \Gamma_1^-$
Γ_7^-	2	−2	$-\sqrt{2}$	$\sqrt{2}$	0	0	0	−2	2	$\sqrt{2}$	$-\sqrt{2}$	0	0	0	c	$\Gamma_6^+ \times \Gamma_3^-$

Table 41. Compatibility Table

D_{4h}	Γ_1^+	Γ_2^+	Γ_3^+	Γ_4^+	Γ_5^+	Γ_1^-	Γ_2^-	Γ_3^-	Γ_4^-	Γ_5^-	Γ_6^+	Γ_7^+	Γ_6^-	Γ_7^-
D_4	Γ_1	Γ_2	Γ_3	Γ_4	Γ_5	Γ_1	Γ_2	Γ_3	Γ_4	Γ_5	Γ_6	Γ_7	Γ_6	Γ_7
C_{4v} :E(z)	Γ_1	Γ_2	Γ_3	Γ_4	Γ_5	Γ_2	Γ_1	Γ_4	Γ_3	Γ_5	Γ_6	Γ_7	Γ_6	Γ_7
D_{2d}	Γ_1	Γ_2	Γ_3	Γ_4	Γ_5	Γ_3	Γ_4	Γ_1	Γ_2	Γ_5	Γ_6	Γ_7	Γ_7	Γ_6
C_{4h} :H(z)	Γ_1^+	Γ_1^+	Γ_2^+	Γ_2^+	$\Gamma_3^+ + \Gamma_4^+$	Γ_1^-	Γ_1^-	Γ_2^-	Γ_2^-	$\Gamma_3^- + \Gamma_4^-$	$\Gamma_5^+ + \Gamma_6^+$	$\Gamma_7^+ + \Gamma_8^+$	$\Gamma_5^- + \Gamma_6^-$	$\Gamma_7^- + \Gamma$
C_{2h} :H(x)	Γ_1^+	Γ_2^+	Γ_1^+	Γ_2^+	$\Gamma_1^+ + \Gamma_2^+$	Γ_1^-	Γ_2^-	Γ_1^-	Γ_2^-	$\Gamma_1^- + \Gamma_2^-$	$\Gamma_3^+ + \Gamma_4^+$	$\Gamma_3^+ + \Gamma_4^+$	$\Gamma_3^- + \Gamma_4^-$	$\Gamma_3^- + \Gamma$
D_{2h}	Γ_1^+	Γ_3^+	Γ_1^+	Γ_3^+	$\Gamma_2^+ + \Gamma_4^+$	Γ_1^-	Γ_3^-	Γ_1^-	Γ_3^-	$\Gamma_2^- + \Gamma_4^-$	Γ_5^+	Γ_5^+	Γ_5^-	Γ_5^-
C_{2v} :E(x)	Γ_1	Γ_2	Γ_1	Γ_2	$\Gamma_3 + \Gamma_4$	Γ_3	Γ_4	Γ_3	Γ_4	$\Gamma_1 + \Gamma_2$	Γ_5	Γ_5	Γ_5	Γ_5

(16) C_3. This is a cyclic group consisting of the identity; a rotation through 120°, C_3; and a rotation through −120°, C_3^{-1}.

Table 42. Character Table and Basis Functions for the Group C_3

C_3	E	$\bar{E}$	C_3	$\bar{C}_3$	C_3^{-1}	$\bar{C}_3^{-1}$	Time Inv.	Bases
Γ_1	1	1	1	1	1	1	a	z or S_z
Γ_2	1	1	ω^2	ω^2	$-\omega$	$-\omega$	b	$-i(x + iy)$ or $-(S_x + iS_y)$
Γ_3	1	1	$-\omega$	$-\omega$	ω^2	ω^2	b	$i(x - iy)$ or $(S_x - iS_y)$
Γ_4	1	−1	ω	$-\omega$	$-\omega^2$	ω^2	b	$\phi(1/2, 1/2)$
Γ_5	1	−1	$-\omega^2$	ω^2	ω	$-\omega$	b	$\phi(1/2, -1/2)$
Γ_6	1	−1	−1	1	−1	1	a	$\phi(3/2,3/2)$ or $\phi(3/2,-3/2)$

$\omega = \exp(\pi i/3)$

Table 43. Multiplication Table for the Group C_3

Γ_1	Γ_2	Γ_3	Γ_4	Γ_5	Γ_6	
Γ_1	Γ_2	Γ_3	Γ_4	Γ_5	Γ_6	Γ_1
	Γ_3	Γ_1	Γ_6	Γ_4	Γ_5	Γ_2
		Γ_2	Γ_5	Γ_6	Γ_4	Γ_3
			Γ_2	Γ_1	Γ_3	Γ_4
				Γ_3	Γ_2	Γ_5
					Γ_1	Γ_6

Table 44. Coupling Coefficients for the Group C_3

$$u_3 = \mathbf{K}u_2;\ u_2 = \mathbf{K}u_3$$
$$u_5 = \mathbf{K}u_4;\ u_4 = -\mathbf{K}u_5$$

$u_2v_2 = \psi_3$ $u_3v_3 = \psi_2$ $u_4v_4 = \psi_2$

$u_2v_5 = \psi_4$ $u_3v_4 = -\psi_5$ $u_5v_5 = \psi_3$

	u_2v_3	u_3v_2
ψ_1	$1/\sqrt{2}$	$1/\sqrt{2}$
ψ_1	$i/\sqrt{2}$	$-i\sqrt{2}$

	u_2v_4	u_3v_5
ψ_6	1	0
$\mathbf{K}\psi_6$	0	1

	u_2v_6	$u_2\mathbf{K}v_6$
ψ_5	0	1
ψ_5	−1	0

	u_3v_6	$u_3\mathbf{K}v_6$
ψ_4	1	0
ψ_4	0	1

	u_4v_5	u_5v_4
ψ_1	$1/\sqrt{2}$	$-1/\sqrt{2}$
ψ_1	$i/\sqrt{2}$	$i/\sqrt{2}$

	u_4v_6	$u_4\mathbf{K}v_6$
ψ_3	0	−1
ψ_3	1	0

	u_5v_6	$u_5\mathbf{K}v_6$
ψ_2	1	0
ψ_2	0	1

	u_6v_6	$u_6\mathbf{K}v_6$	$\mathbf{K}u_6v_6$	$\mathbf{K}u_6\mathbf{K}v_6$
ψ_1	1	0	0	0
ψ_1	0	0	0	1
ψ_1	0	1	0	0
ψ_1	0	0	−1	0

Table 45. Compatibility Table

C_3:H(z):E(z)	Γ_1	Γ_2	Γ_3	Γ_4	Γ_5	Γ_6
C_1	Γ_1	Γ_1	Γ_1	Γ_2	Γ_2	Γ_2

Table 46. Full Rotation Group Compatibility Table for the Group C_3

$D_0^{\pm}$	Γ_1
$D_1^{\pm}$	$\Gamma_1 + \Gamma_2 + \Gamma_3$
$D_2^{\pm}$	$\Gamma_1 + 2\Gamma_2 + 2\Gamma_3$
$D_3^{\pm}$	$3\Gamma_1 + 2\Gamma_2 + 2\Gamma_3$
$D_4^{\pm}$	$3\Gamma_1 + 3\Gamma_2 + 3\Gamma_3$
$D_5^{\pm}$	$3\Gamma_1 + 4\Gamma_2 + 4\Gamma_3$
$D_6^{\pm}$	$5\Gamma_1 + 4\Gamma_2 + 4\Gamma_3$
$D_{1/2}^{\pm}$	$\Gamma_4 + \Gamma_5$
$D_{3/2}^{\pm}$	$\Gamma_4 + \Gamma_5 + 2\Gamma_6$
$D_{5/2}^{\pm}$	$2\Gamma_4 + 2\Gamma_5 + 2\Gamma_6$
$D_{7/2}^{\pm}$	$3\Gamma_4 + 3\Gamma_5 + 2\Gamma_6$
$D_{9/2}^{\pm}$	$3\Gamma_4 + 3\Gamma_5 + 4\Gamma_6$
$D_{11/2}^{\pm}$	$4\Gamma_4 + 4\Gamma_5 + 4\Gamma_6$
$D_{13/2}^{\pm}$	$5\Gamma_4 + 5\Gamma_5 + 4\Gamma_6$

(17) C_{3i}. This group is the direct product of C_3 and C_i ($C_{3i} = C_3 \times C_i$). In addition to the operators of C_3 and C_i it contains the operators $S_6 = IC_3^{-1}$ and $S_6^{-1} = IC_3$.

Table 47. Character Table and Basis Functions for the Group C_{3i}

C_{3i}	E	$\bar{E}$	C_3	$\bar{C}_3$	C_3^{-1}	$\bar{C}_3^{-1}$	I	$\bar{I}$	S_6^{-1}	$\bar{S}_6^{-1}$	S_6	$\bar{S}_6$	Time Inv.	Bases
Γ_1^+	1	1	1	1	1	1	1	1	1	1	1	1	a	R or S_z
Γ_2^+	1	1	ω^2	ω^2	$-\omega$	$-\omega$	1	1	ω^2	ω^2	$-\omega$	$-\omega$	b	$-(S_x + iS_y)$
Γ_3^+	1	1	$-\omega$	$-\omega$	ω^2	ω^2	1	1	$-\omega$	$-\omega$	ω^2	ω^2	b	$(S_x - iS_y)$
Γ_1^-	1	1	1	1	1	1	−1	−1	−1	−1	−1	−1	a	z
Γ_2^-	1	1	ω^2	ω^2	$-\omega$	$-\omega$	−1	−1	$-\omega^2$	$-\omega^2$	ω	ω	b	$-i(x + iy)$
Γ_3^-	1	1	$-\omega$	$-\omega$	ω^2	ω^2	−1	−1	ω	ω	$-\omega^2$	$-\omega^2$	b	$i(x - iy)$
Γ_4^+	1	−1	ω	$-\omega$	$-\omega^2$	ω^2	1	−1	ω	$-\omega$	$-\omega^2$	ω^2	b	$\phi(1/2,1/2)$
Γ_5^+	1	−1	$-\omega^2$	ω^2	ω	$-\omega$	1	−1	$-\omega^2$	ω^2	ω	$-\omega$	b	$\phi(1/2,-1/2)$
Γ_6^+	1	−1	−1	1	−1	1	1	−1	−1	1	−1	1	a	$\phi(3/2,3/2)$ or $\phi(3/2,-3/2)$
Γ_4^-	1	−1	ω	$-\omega$	$-\omega^2$	ω^2	−1	1	$-\omega$	ω	ω^2	$-\omega^2$	b	$\Gamma_4^+ \times \Gamma_1^-$
Γ_5^-	1	−1	$-\omega^2$	ω^2	ω	$-\omega$	−1	1	ω^2	$-\omega^2$	$-\omega$	ω	b	$\Gamma_5^+ \times \Gamma_1^-$
Γ_6^-	1	−1	−1	1	−1	1	−1	1	1	−1	1	−1	a	$\Gamma_6^+ \times \Gamma_1^-$

$\omega = \exp(\pi i/3)$

Table 48. Compatibility Table

C_{3i}: H(z)	Γ_1^+	Γ_2^+	Γ_3^+	Γ_1^-	Γ_2^-	Γ_3^-	Γ_4^+	Γ_5^+	Γ_6^+	Γ_4^-	Γ_5^-	Γ_6^-
C_3 : E(z)	Γ_1	Γ_2	Γ_3	Γ_1	Γ_2	Γ_3	Γ_4	Γ_5	Γ_6	Γ_4	Γ_5	Γ_6
C_i: H(x)	Γ_1^+	Γ_1^+	Γ_1^+	Γ_1^-	Γ_1^-	Γ_1^-	Γ_2^+	Γ_2^+	Γ_2^+	Γ_2^-	Γ_2^-	Γ_2^-

(18) D_3. This group has a threefold axis (taken as z in Figure 3) and three twofold axes making 120° angles with respect to one another in a plane perpendicular to the threefold axis. We may take these axes to be y, y′, and y″ in Figure 3.

(19) C_{3v}. This group, isomorphic to D_3, has a threefold axis z and three reflections in planes containing the threefold axis and making 120° angles with respect to one another. We may take these planes to be xz, x′z, and x″z in Figure 3.

Table 49. Character Table and Basis Functions for the Groups D_3 and C_{3v}

D_3	E	$\bar{E}$	$2C_3$	$2\bar{C}_3$	$3C_2'$	$3\bar{C}_2'$	Time Inv.	Bases for D_3	Bases for C_{3v}
C_{3v}	E	$\bar{E}$	$2C_3$	$2\bar{C}_3$	$3\sigma_v$	$3\bar{\sigma}_v$			
Γ_1	1	1	1	1	1	1	a	R	R or z
Γ_2	1	1	1	1	−1	−1	a	S_z or z	S_z
Γ_3	2	2	−1	−1	0	0	a	$(S_x - iS_y)$, $-(S_x + iS_y)$	$(S_x - iS_y)$, $-(S_x + iS_y)$
Γ_4	2	−2	1	−1	0	0	c	$\phi(1/2,-1/2)$, $\phi(1/2,1/2)$	$\phi(1/2,-1/2)$, $\phi(1/2,1/2)$
Γ_5	1	−1	−1	1	i	−i	b	$\phi(3/2,-3/2) - i\phi(3/2,3/2)$	$\phi(3/2,-3/2) - i\phi(3/2,3/2)$
Γ_6	1	−1	−1	1	−i	i	b	$-(\phi(3/2,3/2) + i\phi(3/2,-3/2))$	$-(\phi(3/2,3/2) + i\phi(3/2,-3/2))$

Table 50. Multiplication Table for the Groups D_3 and C_{3v}

Γ_1	Γ_2	Γ_3	Γ_4	Γ_5	Γ_6	
Γ_1	Γ_2	Γ_3	Γ_4	Γ_5	Γ_6	Γ_1
	Γ_1	Γ_3	Γ_4	Γ_6	Γ_5	Γ_2
		$\Gamma_1 + \Gamma_2 + \Gamma_3$	$\Gamma_4 + \Gamma_5 + \Gamma_6$	Γ_4	Γ_4	Γ_3
			$\Gamma_1 + \Gamma_2 + \Gamma_3$	Γ_3	Γ_3	Γ_4
				Γ_2	Γ_1	Γ_5
					Γ_2	Γ_6

Table 51. Coupling Coefficients for the Groups D_3 and C_{3v}

$$u_6 = \mathbf{K}u_5; \quad u_5 = -\mathbf{K}u_6$$

$$u_2v_5 = -\psi_6 \qquad u_2v_6 = \psi_5$$

	$u_2v^3_{-1}$	$u_2v^3_1$
ψ^3_{-1}	i	0
ψ^3_1	0	−i

	$u_2v^4_{-1/2}$	$u_2v^4_{1/2}$
$\psi^4_{-1/2}$	i	0
$\psi^4_{1/2}$	0	−i

	$u^3_{-1}v^3_{-1}$	$u^3_{-1}v^3_1$	$u^3_1v^3_{-1}$	$u^3_1v^3_1$
ψ_1	0	$1/\sqrt{2}$	$1/\sqrt{2}$	0
ψ_2	0	$i/\sqrt{2}$	$-i/\sqrt{2}$	0
ψ^3_{-1}	0	0	0	1
ψ^3_1	1	0	0	0

	$u^3_{-1}v^4_{-1/2}$	$u^3_{-1}v^4_{1/2}$	$u^3_1v^4_{-1/2}$	$u^3_1v^4_{1/2}$
$\psi^4_{-1/2}$	0	1	0	0
$\psi^4_{1/2}$	0	0	−1	0
ψ_5	$1/\sqrt{2}$	0	0	$-i/\sqrt{2}$
ψ_6	$i/\sqrt{2}$	0	0	$-1/\sqrt{2}$

	$u^3_{-1}v_5$	$u^3_{-1}v_6$	$u^3_1v_5$	$u^3_1v_6$
$\psi^4_{-1/2}$	0	0	$i/\sqrt{2}$	$-1/\sqrt{2}$
$\psi^4_{1/2}$	$-1/\sqrt{2}$	$i/\sqrt{2}$	0	0
$\psi^4_{-1/2}$	0	0	$1/\sqrt{2}$	$-i/\sqrt{2}$
$\psi^4_{1/2}$	$i/\sqrt{2}$	$-1/\sqrt{2}$	0	0

	$u^4_{-1/2}v^4_{-1/2}$	$u^4_{-1/2}v^4_{1/2}$	$u^4_{1/2}v^4_{-1/2}$	$u^4_{1/2}v^4_{1/2}$
ψ_1	0	$-1/\sqrt{2}$	$1/\sqrt{2}$	0
ψ_2	0	$-i/\sqrt{2}$	$-i/\sqrt{2}$	0
ψ^3_{-1}	1	0	0	0
ψ^3_1	0	0	0	1

	$u^4_{-1/2}v_5$	$u^4_{-1/2}v_6$	$u^4_{1/2}v_5$	$u^4_{1/2}v_6$
ψ^3_{-1}	0	0	$i/\sqrt{2}$	$-1/\sqrt{2}$
ψ^3_1	$1/\sqrt{2}$	$-i/\sqrt{2}$	0	0
ψ^3_{-1}	0	0	$-1/\sqrt{2}$	$i/\sqrt{2}$
ψ^3_1	$i/\sqrt{2}$	$-1/\sqrt{2}$	0	0

	u_5v_5	u_6v_6
ψ_2	$1/\sqrt{2}$	$1/\sqrt{2}$
ψ_2	$i/\sqrt{2}$	$-i/\sqrt{2}$

	u_5v_6	u_6v_5
ψ_1	$1/\sqrt{2}$	$-1/\sqrt{2}$
ψ_1	$i/\sqrt{2}$	$i/\sqrt{2}$

Table 52. Compatibility Table

D_3	C_{3v}: E(z)	Γ_1	Γ_2	Γ_3	Γ_4	Γ_5	Γ_6
C_3 :H(z): E(z)	C_3: H(z)	Γ_1	Γ_1	$\Gamma_2 + \Gamma_3$	$\Gamma_4 + \Gamma_5$	Γ_6	Γ_6
C_2 :H(y): E(y)	C_s: E(x) : H(y)	Γ_1	Γ_2	$\Gamma_1 + \Gamma_2$	$\Gamma_3 + \Gamma_4$	Γ_3	Γ_4

Table 53. Full Rotation Group Compatibility Table for the Group D_3

$D_0^{\pm}$	Γ_1
$D_1^{\pm}$	$\Gamma_2 + \Gamma_3$
$D_2^{\pm}$	$\Gamma_1 + 2\Gamma_3$
$D_3^{\pm}$	$\Gamma_1 + 2\Gamma_2 + 2\Gamma_3$
$D_4^{\pm}$	$2\Gamma_1 + \Gamma_2 + 3\Gamma_3$
$D_5^{\pm}$	$\Gamma_1 + 2\Gamma_2 + 4\Gamma_3$
$D_6^{\pm}$	$3\Gamma_1 + 2\Gamma_2 + 4\Gamma_3$
$D_{1/2}^{\pm}$	Γ_4
$D_{3/2}^{\pm}$	$\Gamma_4 + \Gamma_5 + \Gamma_6$
$D_{5/2}^{\pm}$	$2\Gamma_4 + \Gamma_5 + \Gamma_6$
$D_{7/2}^{\pm}$	$3\Gamma_4 + \Gamma_5 + \Gamma_6$
$D_{9/2}^{\pm}$	$3\Gamma_4 + 2\Gamma_5 + 2\Gamma_6$
$D_{11/2}^{\pm}$	$4\Gamma_4 + 2\Gamma_5 + 2\Gamma_6$
$D_{13/2}^{\pm}$	$5\Gamma_4 + 2\Gamma_5 + 2\Gamma_6$

Table 54. Full Rotation Group Compatibility Table for the Group C_{3v}

D_0^+	Γ_1	D_0^-	Γ_2
D_1^+	$\Gamma_2 + \Gamma_3$	D_1^-	$\Gamma_1 + \Gamma_3$
D_2^+	$\Gamma_1 + 2\Gamma_3$	D_2^-	$\Gamma_2 + 2\Gamma_3$
D_3^+	$\Gamma_1 + 2\Gamma_2 + 2\Gamma_3$	D_3^-	$2\Gamma_1 + \Gamma_2 + 2\Gamma_3$
D_4^+	$2\Gamma_1 + \Gamma_2 + 3\Gamma_3$	D_4^-	$\Gamma_1 + 2\Gamma_2 + 3\Gamma_3$
D_5^+	$\Gamma_1 + 2\Gamma_2 + 4\Gamma_3$	D_5^-	$2\Gamma_1 + \Gamma_2 + 4\Gamma_3$
D_6^+	$3\Gamma_1 + 2\Gamma_2 + 4\Gamma_3$	D_6^-	$2\Gamma_1 + 3\Gamma_2 + 4\Gamma_3$
$D_{1/2}^+$	Γ_4	$D_{1/2}^-$	Γ_4
$D_{3/2}^+$	$\Gamma_4 + \Gamma_5 + \Gamma_6$	$D_{3/2}^-$	$\Gamma_4 + \Gamma_5 + \Gamma_6$
$D_{5/2}^+$	$2\Gamma_4 + \Gamma_5 + \Gamma_6$	$D_{5/2}^-$	$2\Gamma_4 + \Gamma_5 + \Gamma_6$
$D_{7/2}^+$	$3\Gamma_4 + \Gamma_5 + \Gamma_6$	$D_{7/2}^-$	$3\Gamma_4 + \Gamma_5 + \Gamma_6$
$D_{9/2}^+$	$3\Gamma_4 + 2\Gamma_5 + 2\Gamma_6$	$D_{9/2}^-$	$3\Gamma_4 + 2\Gamma_5 + 2\Gamma_6$
$D_{11/2}^+$	$4\Gamma_4 + 2\Gamma_5 + 2\Gamma_6$	$D_{11/2}^-$	$4\Gamma_4 + 2\Gamma_5 + 2\Gamma_6$
$D_{13/2}^+$	$5\Gamma_4 + 2\Gamma_5 + 2\Gamma_6$	$D_{13/2}^-$	$5\Gamma_4 + 2\Gamma_5 + 2\Gamma_6$

(20) D_{3d}. This group is the direct product of D_3 and C_i ($D_{3d} = D_3 \times C_i$). It can also be considered as the direct product of C_{3v} and C_i. It contains all the operations of D_3, C_{3v}, and C_i, and in addition the operations $S_6 = IC_3^{-1}$ and $S_6^{-1} = IC_3$. In Figure 3 we may again take the threefold axis as z. The three twofold axes perpendicular to this and making 120° angles with respect to one another may be taken as y, y′, and y″. The three reflection planes σ_d contain the z axis, make 120° angles with respect to each other, and are each perpendicular to one of the twofold axes. These planes may be taken as xz, x′z, and x″z.

Table 55. Character Table and Basis Functions for the Group D_{3d}

D_{3d}	E	$\bar{E}$	$2C_3$	$2\bar{C}_3$	$3C_2'$	$3\bar{C}_2'$	I	$\bar{I}$	$2S_6$	$2\bar{S}_6$	$3\sigma_d$	$3\bar{\sigma}_d$	Time Inv.	Bases
Γ_1^+	1	1	1	1	1	1	1	1	1	1	1	1	a	R
Γ_2^+	1	1	1	1	−1	−1	1	1	1	1	−1	−1	a	S_z
Γ_3^+	2	2	−1	−1	0	0	2	2	−1	−1	0	0	a	$(S_x - iS_y)$, $-(S_x + iS_y)$
Γ_1^-	1	1	1	1	1	1	−1	−1	−1	−1	−1	−1	a	zS_z
Γ_2^-	1	1	1	1	−1	−1	−1	−1	−1	−1	1	1	a	z
Γ_3^-	2	2	−1	−1	0	0	−2	−2	1	1	0	0	a	$(x - iy)$, $-(x + iy)$
Γ_4^+	2	−2	1	−1	0	0	2	−2	1	−1	0	0	c	$\phi(1/2,-1/2)$, $\phi(1/2, 1/2)$
Γ_5^+	1	−1	−1	1	i	−i	1	−1	−1	1	i	−i	b	$\phi(3/2,-3/2)$ $-i\phi(3/2,3/2)$
Γ_6^+	1	−1	−1	1	−i	i	1	−1	−1	1	−i	i	b	$-(\phi(3/2,3/2)$ $-i\phi(3/2,-3/2))$
Γ_4^-	2	−2	1	−1	0	0	−2	2	−1	1	0	0	c	$\Gamma_4^+ \times \Gamma_1^-$
Γ_5^-	1	−1	−1	1	i	−i	−1	1	1	−1	−i	i	b	$\Gamma_5^+ \times \Gamma_1^-$
Γ_6^-	1	−1	−1	1	−i	i	−1	1	1	−1	i	−i	b	$\Gamma_6^+ \times \Gamma_1^-$

Table 56. Compatibility Table

D_{3d}	Γ_1^+	Γ_2^+	Γ_3^+	Γ_1^-	Γ_2^-	Γ_3^-	Γ_4^+	Γ_5^+	Γ_6^+	Γ_4^-	Γ_5^-	Γ_6^-
D_3	Γ_1	Γ_2	Γ_3	Γ_1	Γ_2	Γ_3	Γ_4	Γ_5	Γ_6	Γ_4	Γ_5	Γ_6
C_{3i}:H(z)	Γ_1^+	Γ_1^+	$\Gamma_2^+ + \Gamma_3^+$	Γ_1^-	Γ_1^-	$\Gamma_2^- + \Gamma_3^-$	$\Gamma_4^+ + \Gamma_5^+$	Γ_6^+	Γ_6^+	$\Gamma_4^- + \Gamma_5^-$	Γ_6^-	Γ_6^-
C_{3v}:E(z)	Γ_1	Γ_2	Γ_3	Γ_2	Γ_1	Γ_3	Γ_4	Γ_5	Γ_6	Γ_5	Γ_4	Γ_6
C_{2h}:H(y)	Γ_1^+	Γ_2^+	$\Gamma_1^+ + \Gamma_2^+$	Γ_1^-	Γ_2^-	$\Gamma_1^- + \Gamma_2^-$	$\Gamma_3^+ + \Gamma_4^+$	Γ_3^+	Γ_4^+	$\Gamma_3^- + \Gamma_4^-$	Γ_3^-	Γ_4^-
C_2:E(y)	Γ_1	Γ_2	$\Gamma_1 + \Gamma_2$	Γ_1	Γ_2	$\Gamma_1 + \Gamma_2$	$\Gamma_3 + \Gamma_4$	Γ_3	Γ_4	$\Gamma_3 + \Gamma_4$	Γ_3	Γ_4

(21) C_6. This is a cyclic group consisting of the identity; a rotation through 60°, C_6; a rotation through −60°, C_6^{-1}; a rotation through 120°, C_3; a rotation through −120°, C_3^{-1}; and a rotation through 180°, C_2. This group may be considered as the direct product of C_3 and C_2.

(22) C_{3h}. This group, isomorphic to C_6, consists of a threefold axis z, and a reflection plane perpendicular to the axis xy. In addition to the operations of C_3 the group contains the reflection σ_h and the operations S_3 and S_3^{-1}. This group may be considered as the direct product of C_3 and C_s. ($C_{3h} = C_3 \times C_s$).

Table 57. Character Table and Basis Functions for the Groups C_6 and C_{3h}

C_6	E	$\bar{E}$	C_6	$\bar{C}_6$	C_3	$\bar{C}_3$	C_2	$\bar{C}_2$	C_3^{-1}	$\bar{C}_3^{-1}$	C_6^{-1}	$\bar{C}_6^{-1}$	Time	Bases	Bases
C_{3h}	E	$\bar{E}$	S_3^{-1}	$\bar{S}_3^{-1}$	C_3	$\bar{C}_3$	σ_h	$\bar{\sigma}_h$	C_3^{-1}	$\bar{C}_3^{-1}$	S_3	$\bar{S}_3$	Inv.	for C_6	for C_{3h}
Γ_1	1	1	1	1	1	1	1	1	1	1	1	1	a	R or z or S_z	R
Γ_2	1	1	$-\omega^2$	$-\omega^2$	ω^4	ω^4	1	1	$-\omega^2$	$-\omega^2$	ω^4	ω^4	b	$(x - iy)^2$	$(x - iy)^2$
Γ_3	1	1	ω^4	ω^4	$-\omega^2$	$-\omega^2$	1	1	ω^4	ω^4	$-\omega^2$	$-\omega^2$	b	$(x + iy)^2$	$(x + iy)^2$
Γ_4	1	1	−1	−1	1	1	−1	−1	1	1	−1	−1	a	$(x \pm iy)^3$	z
Γ_5	1	1	ω^2	ω^2	ω^4	ω^4	−1	−1	$-\omega^2$	$-\omega^2$	$-\omega^4$	$-\omega^4$	b	$-(S_x + iS_y)$	$-(S_x + iS_y)$
Γ_6	1	1	$-\omega^4$	$-\omega^4$	$-\omega^2$	$-\omega^2$	−1	−1	ω^4	ω^4	ω^2	ω^2	b	$(S_x - iS_y)$	$(S_x - iS_y)$
Γ_7	1	−1	ω	$-\omega$	ω^2	$-\omega^2$	i	−i	$-\omega^4$	ω^4	$-\omega^5$	ω^5	b	$\phi(1/2,1/2)$	$\phi(1/2,1/2)$
Γ_8	1	−1	$-\omega^5$	ω^5	$-\omega^4$	ω^4	−i	i	ω^2	$-\omega^2$	ω	$-\omega$	b	$\phi(1/2,-1/2)$	$\phi(1/2,-1/2)$
Γ_9	1	−1	$-\omega$	ω	ω^2	$-\omega^2$	−i	i	$-\omega^4$	ω^4	ω^5	$-\omega^5$	b	$\phi(5/2,-5/2)$	$\phi(5/2,-5/2)$
Γ_{10}	1	−1	ω^5	$-\omega^5$	$-\omega^4$	ω^4	i	−i	ω^2	$-\omega^2$	$-\omega$	ω	b	$\phi(5/2,5/2)$	$\phi(5/2,5/2)$
Γ_{11}	1	−1	−i	i	−1	1	i	−i	−1	1	i	−i	b	$\phi(3/2,-3/2)$	$\phi(3/2,-3/2)$
Γ_{12}	1	−1	i	−i	−1	1	−i	i	−1	1	−i	i	b	$\phi(3/2,3/2)$	$\phi(3/2,3/2)$

$\omega = \exp(\pi i/6)$

Table 58. Multiplication Table for the Groups C_6 and C_{3h}

Γ_1	Γ_2	Γ_3	Γ_4	Γ_5	Γ_6	Γ_7	Γ_8	Γ_9	Γ_{10}	Γ_{11}	Γ_{12}	
Γ_1	Γ_2	Γ_3	Γ_4	Γ_5	Γ_6	Γ_7	Γ_8	Γ_9	Γ_{10}	Γ_{11}	Γ_{12}	Γ_1
	Γ_3	Γ_1	Γ_5	Γ_6	Γ_4	Γ_{11}	Γ_9	Γ_{12}	Γ_7	Γ_{10}	Γ_8	Γ_2
		Γ_2	Γ_6	Γ_4	Γ_5	Γ_{10}	Γ_{12}	Γ_8	Γ_{11}	Γ_7	Γ_9	Γ_3
			Γ_1	Γ_2	Γ_3	Γ_9	Γ_{10}	Γ_7	Γ_8	Γ_{12}	Γ_{11}	Γ_4
				Γ_3	Γ_1	Γ_{12}	Γ_7	Γ_{11}	Γ_9	Γ_8	Γ_{10}	Γ_5
					Γ_2	Γ_8	Γ_{11}	Γ_{10}	Γ_{12}	Γ_9	Γ_7	Γ_6
						Γ_5	Γ_1	Γ_2	Γ_4	Γ_6	Γ_3	Γ_7
							Γ_6	Γ_4	Γ_3	Γ_2	Γ_5	Γ_8
								Γ_5	Γ_1	Γ_3	Γ_6	Γ_9
									Γ_6	Γ_5	Γ_2	Γ_{10}
										Γ_4	Γ_1	Γ_{11}
											Γ_4	Γ_{12}

Table 59. Coupling Coefficients for the Groups C_6 and C_{3h}

$u_3 = \mathbf{K}u_2;\ u_2 = \mathbf{K}u_3$ $\qquad$ $u_8 = \mathbf{K}u_7;\ u_7 = -\mathbf{K}u_8$

$u_6 = \mathbf{K}u_5;\ u_5 = \mathbf{K}u_6$ $\qquad$ $u_{10} = \mathbf{K}u_9;\ u_9 = -\mathbf{K}u_{10}$

$u_{11} = \mathbf{K}u_{12};u_{12} = -\mathbf{K}u_{11}$

$u_2v_2 = \psi_3$	$u_3v_3 = \psi_2$	$u_4v_4 = \psi_1$
$u_2v_4 = \psi_5$	$u_3v_4 = \psi_6$	$u_4v_5 = \psi_2$
$u_2v_5 = \psi_6$	$u_3v_6 = \psi_5$	$u_4v_6 = \psi_3$
$u_2v_7 = -\psi_{11}$	$u_3v_7 = -\psi_{10}$	$u_4v_7 = -\psi_8$
$u_2v_8 = \psi_9$	$u_3v_8 = \psi_{12}$	$u_4v_8 = \psi_7$
$u_2v_9 = \psi_{12}$	$u_3v_9 = -\psi_8$	$u_4v_9 = -\psi_{10}$
$u_2v_{10} = \psi_7$	$u_3v_{10} = \psi_{11}$	$u_4v_{10} = \psi_9$
$u_2v_{11} = \psi_{10}$	$u_3v_{11} = \psi_7$	$u_4v_{11} = -\psi_{12}$
$u_2v_{12} = -\psi_8$	$u_3v_{12} = \psi_9$	$u_4v_{12} = \psi_{11}$

Table 59. Coupling Coefficients for the Groups C_6 and C_{3h} (Cont.)

$u_5v_5 = \psi_3$ | $u_6v_6 = \psi_2$
$u_5v_7 = \psi_{12}$ | $u_6v_7 = -\psi_8$
$u_5v_8 = \psi_7$ | $u_6v_8 = \psi_{11}$
$u_5v_9 = -\psi_{11}$ | $u_6v_9 = -\psi_{10}$
$u_5v_{10} = \psi_9$ | $u_6v_{10} = \psi_{12}$
$u_5v_{11} = \psi_8$ | $u_6v_{11} = \psi_9$
$u_5v_{12} = -\psi_{10}$ | $u_6v_{12} = \psi_7$

$u_7v_7 = \psi_5$ | $u_8v_8 = \psi_6$
$u_7v_9 = -\psi_2$ | $u_8v_{10} = -\psi_3$
$u_7v_{11} = \psi_6$ | $u_8v_{11} = \psi_2$
$u_7v_{12} = \psi_3$ | $u_8v_{12} = -\psi_5$

$u_9v_9 = \psi_5$ | $u_{10}v_{10} = \psi_6$
$u_9v_{11} = \psi_3$ | $u_{10}v_{11} = \psi_5$
$u_9v_{12} = \psi_6$ | $u_{10}v_{12} = -\psi_2$

	u_2v_3	u_3v_2
ψ_1	$1/\sqrt{2}$	$1/\sqrt{2}$
ψ_1	$-i/\sqrt{2}$	$i/\sqrt{2}$

	u_3v_5	u_2v_6
ψ_4	$1/\sqrt{2}$	$1/\sqrt{2}$
ψ_4	$i/\sqrt{2}$	$-i/\sqrt{2}$

	u_5v_6	u_6v_5
ψ_1	$1/\sqrt{2}$	$1/\sqrt{2}$
ψ_1	$-i/\sqrt{2}$	$i/\sqrt{2}$

	u_7v_8	u_8v_7
ψ_1	$1/\sqrt{2}$	$-1/\sqrt{2}$
ψ_1	$i/\sqrt{2}$	$i/\sqrt{2}$

	u_7v_{10}	u_8v_9
ψ_4	$1/\sqrt{2}$	$-1/\sqrt{2}$
ψ_4	$i/\sqrt{2}$	$i/\sqrt{2}$

	u_9v_{10}	$u_{10}v_9$
ψ_1	$1/\sqrt{2}$	$-1/\sqrt{2}$
ψ_1	$i/\sqrt{2}$	$i/\sqrt{2}$

	$u_{11}v_{11}$	$u_{12}v_{12}$
ψ_4	$1/\sqrt{2}$	$1/\sqrt{2}$
ψ_4	$-i/\sqrt{2}$	$i/\sqrt{2}$

	$u_{11}v_{12}$	$u_{12}v_{11}$
ψ_1	$-1/\sqrt{2}$	$1/\sqrt{2}$
ψ_1	$i/\sqrt{2}$	$i/\sqrt{2}$

Table 60. Compatibility Table

C_6:H(z):E(z)	C_{3h}:H_z	Γ_1	Γ_2	Γ_3	Γ_4	Γ_5	Γ_6	Γ_7	Γ_8	Γ_9	Γ_{10}	Γ_{11}	Γ_{12}
C_3	C_3:E(z)	Γ_1	Γ_2	Γ_3	Γ_1	Γ_2	Γ_3	Γ_4	Γ_5	Γ_4	Γ_5	Γ_6	Γ_6
C_2	C_s:E(x)	Γ_1	Γ_1	Γ_1	Γ_2	Γ_2	Γ_2	Γ_3	Γ_4	Γ_4	Γ_3	Γ_3	Γ_4

Table 61. Full Rotation Group Compatibility Table for the Group C_6

$D_0^\pm$	Γ_1
$D_1^\pm$	$\Gamma_1 + \Gamma_5 + \Gamma_6$
$D_2^\pm$	$\Gamma_1 + \Gamma_2 + \Gamma_3 + \Gamma_5 + \Gamma_6$
$D_3^\pm$	$\Gamma_1 + \Gamma_2 + \Gamma_3 + 2\Gamma_4 + \Gamma_5 + \Gamma_6$
$D_4^\pm$	$\Gamma_1 + 2\Gamma_2 + 2\Gamma_3 + 2\Gamma_4 + \Gamma_5 + \Gamma_6$
$D_5^\pm$	$\Gamma_1 + 2\Gamma_2 + 2\Gamma_3 + 2\Gamma_4 + 2\Gamma_5 + 2\Gamma_6$
$D_6^\pm$	$3\Gamma_1 + 2\Gamma_2 + 2\Gamma_3 + 2\Gamma_4 + 2\Gamma_5 + 2\Gamma_6$
$D_{1/2}^\pm$	$\Gamma_7 + \Gamma_8$
$D_{3/2}^\pm$	$\Gamma_7 + \Gamma_8 + \Gamma_{11} + \Gamma_{12}$
$D_{5/2}^\pm$	$\Gamma_7 + \Gamma_8 + \Gamma_9 + \Gamma_{10} + \Gamma_{11} + \Gamma_{12}$
$D_{7/2}^\pm$	$\Gamma_7 + \Gamma_8 + 2\Gamma_9 + 2\Gamma_{10} + \Gamma_{11} + \Gamma_{12}$
$D_{9/2}^\pm$	$\Gamma_7 + \Gamma_8 + 2\Gamma_9 + 2\Gamma_{10} + 2\Gamma_{11} + 2\Gamma_{12}$
$D_{11/2}^\pm$	$2\Gamma_7 + 2\Gamma_8 + 2\Gamma_9 + 2\Gamma_{10} + 2\Gamma_{11} + 2\Gamma_{12}$
$D_{13/2}^\pm$	$3\Gamma_7 + 3\Gamma_8 + 2\Gamma_9 + 2\Gamma_{10} + 2\Gamma_{11} + 2\Gamma_{12}$

Table 62. Full Rotation Group Compatibility Table for the Group C_{3h}

D_0^+	Γ_1	D_0^-	Γ_4
D_1^+	$\Gamma_1 + \Gamma_5 + \Gamma_6$	D_1^-	$\Gamma_2 + \Gamma_3 + \Gamma_4$
D_2^+	$\Gamma_1 + \Gamma_2 + \Gamma_3 + \Gamma_5 + \Gamma_6$	D_2^-	$\Gamma_2 + \Gamma_3 + \Gamma_4 + \Gamma_5 + \Gamma_6$
D_3^+	$\Gamma_1 + \Gamma_2 + \Gamma_3 + 2\Gamma_4 + \Gamma_5 + \Gamma_6$	D_3^-	$2\Gamma_1 + \Gamma_2 + \Gamma_3 + \Gamma_4 + \Gamma_5 + \Gamma_6$
D_4^+	$\Gamma_1 + 2\Gamma_2 + 2\Gamma_3 + 2\Gamma_4 + \Gamma_5 + \Gamma_6$	D_4^-	$2\Gamma_1 + \Gamma_2 + \Gamma_3 + \Gamma_4 + 2\Gamma_5 + 2\Gamma_6$
D_5^+	$\Gamma_1 + 2\Gamma_2 + 2\Gamma_3 + 2\Gamma_4 + 2\Gamma_5 + 2\Gamma_6$	D_5^-	$2\Gamma_1 + 2\Gamma_2 + 2\Gamma_3 + \Gamma_4 + 2\Gamma_5 + 2\Gamma_6$
D_6^+	$3\Gamma_1 + 2\Gamma_2 + 2\Gamma_3 + 2\Gamma_4 + 2\Gamma_5 + 2\Gamma_6$	D_6^-	$2\Gamma_1 + 2\Gamma_2 + 2\Gamma_3 + 3\Gamma_4 + 2\Gamma_5 + 2\Gamma_6$
$D_{1/2}^+$	$\Gamma_7 + \Gamma_8$	$D_{1/2}^-$	$\Gamma_9 + \Gamma_{10}$
$D_{3/2}^+$	$\Gamma_7 + \Gamma_8 + \Gamma_{11} + \Gamma_{12}$	$D_{3/2}^-$	$\Gamma_9 + \Gamma_{10} + \Gamma_{11} + \Gamma_{12}$
$D_{5/2}^+$	$\Gamma_7 + \Gamma_8 + \Gamma_9 + \Gamma_{10} + \Gamma_{11} + \Gamma_{12}$	$D_{5/2}^-$	$\Gamma_7 + \Gamma_8 + \Gamma_9 + \Gamma_{10} + \Gamma_{11} + \Gamma_{12}$
$D_{7/2}^+$	$\Gamma_7 + \Gamma_8 + 2\Gamma_9 + 2\Gamma_{10} + \Gamma_{11} + \Gamma_{12}$	$D_{7/2}^-$	$2\Gamma_7 + 2\Gamma_8 + \Gamma_9 + \Gamma_{10} + \Gamma_{11} + \Gamma_{12}$
$D_{9/2}^+$	$\Gamma_7 + \Gamma_8 + 2\Gamma_9 + 2\Gamma_{10} + 2\Gamma_{11} + 2\Gamma_{12}$	$D_{9/2}^-$	$2\Gamma_7 + 2\Gamma_8 + \Gamma_9 + \Gamma_{10} + 2\Gamma_{11} + 2\Gamma_{12}$
$D_{11/2}^+$	$2\Gamma_7 + 2\Gamma_8 + 2\Gamma_9 + 2\Gamma_{10} + 2\Gamma_{11} + 2\Gamma_{12}$	$D_{11/2}^-$	$2\Gamma_7 + 2\Gamma_8 + 2\Gamma_9 + 2\Gamma_{10} + 2\Gamma_{11} + 2\Gamma_{12}$
$D_{13/2}^+$	$3\Gamma_7 + 3\Gamma_8 + 2\Gamma_9 + 2\Gamma_{10} + 2\Gamma_{11} + 2\Gamma_{12}$	$D_{13/2}^-$	$2\Gamma_7 + 2\Gamma_8 + 2\Gamma_9 + 3\Gamma_{10} + 2\Gamma_{11} + 2\Gamma_{12}$

(23) C_{6h}. This group is the direct product of C_6 and C_i ($C_{6h} = C_6 \times C_i$). It may also be considered as the direct product of C_6 and C_s, of C_{3h} and C_i, of C_{3h} and C_2, of C_{3i} and C_s, or of C_{3i} and C_2. The reflection plane is perpendicular to the sixfold axis. The group contains in addition to the operations of C_6 the inversion, the reflection σ_h and $S_6 = C_3^{-1} \times I$, $S_6^{-1} = C_3 \times I$, $S_3 = C_6^{-1} \times I$, and $S_3^{-1} = C_6 \times I$.

Table 63. Character Table and

C_{6h}	E	$\bar{E}$	C_6	$\bar{C}_6$	C_3	$\bar{C}_3$	C_2	$\bar{C}_2$	C_3^{-1}	$\bar{C}_3^{-1}$	C_6^{-1}	$\bar{C}_6^{-1}$
Γ_1^+	1	1	1	1	1	1	1	1	1	1	1	1
Γ_2^+	1	1	$-\omega^2$	$-\omega^2$	ω^4	ω^4	1	1	$-\omega^2$	$-\omega^2$	ω^4	ω^4
Γ_3^+	1	1	ω^4	ω^4	$-\omega^2$	$-\omega^2$	1	1	ω^4	ω^4	$-\omega^2$	$-\omega^2$
Γ_4^+	1	1	-1	-1	1	1	-1	-1	1	1	-1	-1
Γ_5^+	1	1	ω^2	ω^2	ω^4	ω^4	-1	-1	$-\omega^2$	$-\omega^2$	$-\omega^4$	$-\omega^4$
Γ_6^+	1	1	$-\omega^4$	$-\omega^4$	$-\omega^2$	$-\omega^2$	-1	-1	ω^4	ω^4	ω^2	ω^2
Γ_1^-	1	1	1	1	1	1	1	1	1	1	1	1
Γ_2^-	1	1	$-\omega^2$	$-\omega^2$	ω^4	ω^4	1	1	$-\omega^2$	$-\omega^2$	ω^4	ω^4
Γ_3^-	1	1	ω^4	ω^4	$-\omega^2$	$-\omega^2$	1	1	ω^4	ω^4	$-\omega^2$	$-\omega^2$
Γ_4^-	1	1	-1	-1	1	1	-1	-1	1	1	-1	-1
Γ_5^-	1	1	ω^2	ω^2	ω^4	ω^4	-1	-1	$-\omega^2$	$-\omega^2$	$-\omega^4$	$-\omega^4$
Γ_6^-	1	1	$-\omega^4$	$-\omega^4$	$-\omega^2$	$-\omega^2$	-1	-1	ω^4	ω^4	ω^2	ω^2
Γ_7^+	1	-1	ω	$-\omega$	ω^2	$-\omega^2$	i	$-$i	$-\omega^4$	ω^4	$-\omega^5$	ω^5
Γ_8^+	1	-1	$-\omega^5$	ω^5	$-\omega^4$	ω^4	$-$i	i	ω^2	$-\omega^2$	ω	$-\omega$
Γ_9^+	1	-1	$-\omega$	ω	ω^2	$-\omega^2$	$-$i	i	$-\omega^4$	ω^4	ω^5	$-\omega^5$
Γ_{10}^+	1	-1	ω^5	$-\omega^5$	$-\omega^4$	ω^4	i	$-$i	ω^2	$-\omega^2$	$-\omega$	ω
Γ_{11}^+	1	-1	$-$i	i	-1	1	i	$-$i	-1	1	i	$-$i
Γ_{12}^+	1	-1	i	$-$i	-1	1	$-$i	i	-1	1	$-$i	i
Γ_7^-	1	-1	ω	$-\omega$	ω^2	$-\omega^2$	i	$-$i	$-\omega^4$	ω^4	$-\omega^5$	ω^5
Γ_8^-	1	-1	$-\omega^5$	ω^5	$-\omega^4$	ω^4	$-$i	i	ω^2	$-\omega^2$	ω	$-\omega$
Γ_9^-	1	-1	$-\omega$	ω	ω^2	$-\omega^2$	$-$i	i	$-\omega^4$	ω^4	ω^5	$-\omega^5$
Γ_{10}^-	1	-1	ω^5	$-\omega^5$	$-\omega^4$	ω^4	i	$-$i	ω^2	$-\omega^2$	$-\omega$	ω
Γ_{11}^-	1	-1	$-$i	i	-1	1	i	$-$i	-1	1	i	$-$i
Γ_{12}^-	1	-1	i	$-$i	-1	1	$-$i	i	-1	1	$-$i	i

$\omega = \exp(\pi i/6)$

Table 64. Compatibility

C_{6h}: H(z)	Γ_1^+	Γ_2^+	Γ_3^+	Γ_4^+	Γ_5^+	Γ_6^+	Γ_1^-	Γ_2^-	Γ_3^-	Γ_4^-	Γ_5^-	Γ_6^-
C_6: E(z)	Γ_1	Γ_2	Γ_3	Γ_4	Γ_5	Γ_6	Γ_1	Γ_2	Γ_3	Γ_4	Γ_5	Γ_6
C_{3h}	Γ_1	Γ_2	Γ_3	Γ_4	Γ_5	Γ_6	Γ_4	Γ_5	Γ_6	Γ_1	Γ_2	Γ_3
C_{3i}	Γ_1^+	Γ_2^+	Γ_3^+	Γ_1^+	Γ_2^+	Γ_3^+	Γ_1^-	Γ_2^-	Γ_3^-	Γ_1^-	Γ_2^-	Γ_3^-
C_{2h}	Γ_1^+	Γ_1^+	Γ_1^+	Γ_2^+	Γ_2^+	Γ_2^+	Γ_1^-	Γ_1^-	Γ_1^-	Γ_2^-	Γ_2^-	Γ_2^-

Basis Functions for the Group C_{6h}

C_{6h}	I	$\bar{I}$	S_3^{-1}	$\bar{S}_3^{-1}$	S_6^{-1}	$\bar{S}_6^{-1}$	σ_h	$\bar{\sigma}_h$	S_6	$\bar{S}_6$	S_3	$\bar{S}_3$	Time Inv.	Bases
Γ_1^+	1	1	1	1	1	1	1	1	1	1	1	1	a	R
Γ_2^+	1	1	$-\omega^2$	$-\omega^2$	ω^4	ω^4	1	1	$-\omega^2$	$-\omega^2$	ω^4	ω^4	b	$(x - iy)^2$
Γ_3^+	1	1	ω^4	ω^4	$-\omega^2$	$-\omega^2$	1	1	ω^4	ω^4	$-\omega^2$	$-\omega^2$	b	$(x + iy)^2$
Γ_4^+	1	1	-1	-1	1	1	-1	-1	1	1	-1	-1	a	$\Gamma_1^- \times \Gamma_4^-$
Γ_5^+	1	1	ω^2	ω^2	ω^4	ω^4	-1	-1	$-\omega^2$	$-\omega^2$	$-\omega^4$	$-\omega^4$	b	$-(S_x + iS_y)$
Γ_6^+	1	1	$-\omega^4$	$-\omega^4$	$-\omega^2$	$-\omega^2$	-1	-1	ω^4	ω^4	ω^2	ω^2	b	$(S_x - iS_y)$
Γ_1^-	-1	-1	-1	-1	-1	-1	-1	-1	-1	-1	-1	-1	a	z
Γ_2^-	-1	-1	ω^2	ω^2	$-\omega^4$	$-\omega^4$	-1	-1	ω^2	ω^2	$-\omega^4$	$-\omega^4$	b	$\Gamma_1^- \times \Gamma_2^+$
Γ_3^-	-1	-1	$-\omega^4$	$-\omega^4$	ω^2	ω^2	-1	-1	$-\omega^4$	$-\omega^4$	ω^2	ω^2	b	$\Gamma_1^- \times \Gamma_3^+$
Γ_4^-	-1	-1	1	1	-1	-1	1	1	-1	-1	1	1	a	$(x \pm iy)^3$
Γ_5^-	-1	-1	$-\omega^2$	$-\omega^2$	$-\omega^4$	$-\omega^4$	1	1	ω^2	ω^2	ω^4	ω^4	b	$-(x + iy)$
Γ_6^-	-1	-1	ω^4	ω^4	ω^2	ω^2	1	1	$-\omega^4$	$-\omega^4$	$-\omega^2$	$-\omega^2$	b	$(x - iy)$
Γ_7^+	1	-1	ω	$-\omega$	ω^2	$-\omega^2$	i	$-i$	$-\omega^4$	ω^4	$-\omega^5$	ω^5	b	$\phi(1/2,1/2)$
Γ_8^+	1	-1	$-\omega^5$	ω^5	$-\omega^4$	ω^4	$-i$	i	ω^2	$-\omega^2$	ω	$-\omega$	b	$\phi(1/2,-1/2)$
Γ_9^+	1	-1	$-\omega$	ω	ω^2	$-\omega^2$	$-i$	i	$-\omega^4$	ω^4	ω^5	$-\omega^5$	b	$\phi(5/2,-5/2)$
Γ_{10}^+	1	-1	ω^5	$-\omega^5$	$-\omega^4$	ω^4	i	$-i$	ω^2	$-\omega^2$	$-\omega$	ω	b	$\phi(5/2,5/2)$
Γ_{11}^+	1	-1	$-i$	i	-1	1	i	$-i$	-1	1	i	$-i$	b	$\phi(3/2,-3/2)$
Γ_{12}^+	1	-1	i	$-i$	-1	1	$-i$	i	-1	1	$-i$	i	b	$\phi(3/2,3/2)$
Γ_7^-	-1	1	$-\omega$	ω	$-\omega^2$	ω^2	$-i$	i	ω^4	$-\omega^4$	ω^5	$-\omega^5$	b	$\Gamma_7^+ \times \Gamma_1^-$
Γ_8^-	-1	1	ω^5	$-\omega^5$	ω^4	$-\omega^4$	i	$-i$	$-\omega^2$	ω^2	$-\omega$	ω	b	$\Gamma_8^+ \times \Gamma_1^-$
Γ_9^-	-1	1	ω	$-\omega$	$-\omega^2$	ω^2	i	$-i$	ω^4	$-\omega^4$	$-\omega^5$	ω^5	b	$\Gamma_9^+ \times \Gamma_1^-$
Γ_{10}^-	-1	1	$-\omega^5$	ω^5	ω^4	$-\omega^4$	$-i$	i	$-\omega^2$	ω^2	ω	$-\omega$	b	$\Gamma_{10}^+ \times \Gamma_1^-$
Γ_{11}^-	-1	1	i	$-i$	1	-1	$-i$	i	1	-1	$-i$	i	b	$\Gamma_{11}^+ \times \Gamma_1^-$
Γ_{12}^-	-1	1	$-i$	i	1	-1	i	$-i$	1	-1	i	$-i$	b	$\Gamma_{12}^+ \times \Gamma_1^-$

$$\omega = \exp(\pi i/6)$$

Table

C_{6h}: H(z)	Γ_7^+	Γ_8^+	Γ_9^+	Γ_{10}^+	Γ_{11}^+	Γ_{12}^+	Γ_7^-	Γ_8^-	Γ_9^-	Γ_{10}^-	Γ_{11}^-	Γ_{12}^-
C_6: E(z)	Γ_7	Γ_8	Γ_9	Γ_{10}	Γ_{11}	Γ_{12}	Γ_7	Γ_8	Γ_9	Γ_{10}	Γ_{11}	Γ_{12}
C_{3h}	Γ_7	Γ_8	Γ_9	Γ_{10}	Γ_{11}	Γ_{12}	Γ_9	Γ_{10}	Γ_7	Γ_8	Γ_{12}	Γ_{11}
C_{3i}	Γ_4^+	Γ_5^+	Γ_4^+	Γ_5^+	Γ_6^+	Γ_6^+	Γ_4^-	Γ_5^-	Γ_4^-	Γ_5^-	Γ_6^-	Γ_6^-
C_{2h}	Γ_3^+	Γ_4^+	Γ_4^+	Γ_3^+	Γ_3^+	Γ_4^+	Γ_3^-	Γ_4^-	Γ_4^-	Γ_3^-	Γ_3^-	Γ_4^-

(24) D_6. This group consists of a sixfold axis z and six twofold axes lying in a plane perpendicular to the sixfold axis and making 60° angles with one another. In Figure 3 we may take the axes of the operations $C_{2'}$ to be y, y′, and y″ and those of $C_{2''}$ to be x, x′, and x″. This may be considered as the direct product of D_3 and C_2.

(25) C_{6v}. This group, isomorphic to D_6, consists of a sixfold axis z and six reflection planes intersecting in the sixfold axis and making 60° angles with one another. In Figure 3 we may take the planes of σ_d to be xz, x′z, and x″z and those of σ_v to be yz, y′z, and y″z. This may be considered as the direct product of C_{3v} and C_2.

(26) D_{3h}. This group, isomorphic to D_6, may be considered as the direct product of D_3 and C_s ($D_{3h} = D_3 \times C_s$). In addition to the operations of D_3 and C_s, we have S_3 and S_3^{-1} and three reflections (σ_v) intersecting in the threefold axis and making 120° angles with one another. If the threefold axis in Figure 3 is z and if the three twofold axis of D_3 are y, y′, and y″, the three reflections are in the planes yz, y′z, and y″z.

Table 65. Character Table and Basis Functions for the Groups D_6, C_{6v} and D_{3h}

D_6	E	$\bar{E}$	C_2 $\bar{C}_2$	$2C_3$	$2\bar{C}_3$	$2C_6$	$2\bar{C}_6$	$3C_2'$ $3\bar{C}_2'$	$3C_2''$ $3\bar{C}_2''$				
C_{6v}	E	$\bar{E}$	C_2 $\bar{C}_2$	$2C_3$	$2\bar{C}_3$	$2C_6$	$2\bar{C}_6$	$3\sigma_d$ $3\bar{\sigma}_d$	$3\sigma_v$ $3\bar{\sigma}_v$				
D_{3h}	E	$\bar{E}$	σ_h $\bar{\sigma}_h$	$2C_3$	$2\bar{C}_3$	$2S_3$	$2\bar{S}_3$	$3C_2'$ $3\bar{C}_2'$	$3\sigma_v$ $3\bar{\sigma}_v$	Time Inv.	Bases for D_6	Bases for C_{6v}	Bases for D_{3h}
Γ_1	1	1	1	1	1	1	1	1	1	a	R	R or z	R
Γ_2	1	1	1	1	1	1	1	-1	-1	a	S_z or z	S_z	S_z
Γ_3	1	1	-1	1	1	-1	-1	1	-1	a	$y^3 - 3yx^2$	$x^3 - 3xy^2$	zS_z
Γ_4	1	1	-1	1	1	-1	-1	-1	1	a	$x^3 - 3xy^2$	$y^3 - 3yx^2$	z
Γ_5	2	2	-2	-1	-1	1	1	0	0	a	$(S_x - iS_y)$, $-(S_x + iS_y)$	$(S_x - iS_y)$, $-(S_x + iS_y)$	$\Gamma_3 \times \Gamma_6$
Γ_6	2	2	2	-1	-1	-1	-1	0	0	a	$\Gamma_3 \times \Gamma_5$	$\Gamma_3 \times \Gamma_5$	$(S_x - iS_y)$, $-(S_x + iS_y)$
Γ_7	2	-2	0	1	-1	$\sqrt{3}$	$-\sqrt{3}$	0	0	c	$\phi(1/2,-1/2)$, $\phi(1/2,1/2)$	$\phi(1/2,-1/2)$, $\phi(1/2,1/2)$	$\phi(1/2,-1/2)$, $\phi(1/2,1/2)$
Γ_8	2	-2	0	1	-1	$-\sqrt{3}$	$\sqrt{3}$	0	0	c	$\Gamma_7 \times \Gamma_3$	$\Gamma_7 \times \Gamma_3$	$\Gamma_7 \times \Gamma_3$
Γ_9	2	-2	0	-2	2	0	0	0	0	c	$\phi(3/2,-3/2)$, $\phi(3/2,3/2)$	$\phi(3/2,-3/2)$, $\phi(3/2,3/2)$	$\phi(3/2,-3/2)$, $\phi(3/2,3/2)$

Table 66. Multiplication Table for the Groups D_6, C_{6v} and D_{3h}

Γ_1	Γ_2	Γ_3	Γ_4	Γ_5	Γ_6	Γ_7	Γ_8	Γ_9	
Γ_1	Γ_2	Γ_3	Γ_4	Γ_5	Γ_6	Γ_7	Γ_8	Γ_9	Γ_1
	Γ_1	Γ_4	Γ_3	Γ_5	Γ_6	Γ_7	Γ_8	Γ_9	Γ_2
		Γ_1	Γ_2	Γ_6	Γ_5	Γ_8	Γ_7	Γ_9	Γ_3
			Γ_1	Γ_6	Γ_5	Γ_8	Γ_7	Γ_9	Γ_4
				$\Gamma_1+\Gamma_2+\Gamma_6$	$\Gamma_3+\Gamma_4+\Gamma_5$	$\Gamma_7+\Gamma_9$	$\Gamma_8+\Gamma_9$	$\Gamma_7+\Gamma_8$	Γ_5
					$\Gamma_1+\Gamma_2+\Gamma_6$	$\Gamma_8+\Gamma_9$	$\Gamma_7+\Gamma_9$	$\Gamma_7+\Gamma_8$	Γ_6
						$\Gamma_1+\Gamma_2+\Gamma_5$	$\Gamma_3+\Gamma_4+\Gamma_6$	$\Gamma_5+\Gamma_6$	Γ_7
							$\Gamma_1+\Gamma_2+\Gamma_5$	$\Gamma_5+\Gamma_6$	Γ_8
								$\Gamma_1+\Gamma_2+\Gamma_3+\Gamma_4$	Γ_9

Table 67. Coupling Coefficients for the Groups D_6, C_{6v}, and D_{3h}

	$u_2 v^5_{-1}$	$u_2 v^5_1$
ψ^5_{-1}	i	0
ψ^5_1	0	–i

	$u_2 v^6_{-1}$	$u_2 v^6_1$
ψ^6_{-1}	i	0
ψ^6_1	0	–i

	$u_2 v^7_{-1/2}$	$u_2 v^7_{1/2}$
$\psi^7_{-1/2}$	i	0
$\psi^7_{1/2}$	0	–i

	$u_2 v^8_{-1/2}$	$u_2 v^8_{1/2}$
$\psi^8_{-1/2}$	i	0
$\psi^8_{1/2}$	0	–i

	$u_2 v^9_{-1/2}$	$u_2 v^9_{1/2}$
$\psi^9_{-1/2}$	i	0
$\psi^9_{1/2}$	0	–i

	$u_3 v^5_{-1}$	$u_3 v^5_1$
ψ^6_{-1}	1	0
ψ^6_1	0	1

	$u_3 v^6_{-1}$	$u_3 v^6_1$
ψ^5_{-1}	1	0
ψ^5_1	0	1

	$u_3 v^7_{-1/2}$	$u_3 v^7_{1/2}$
$\psi^8_{-1/2}$	1	0
$\psi^8_{1/2}$	0	1

	$u_3 v^8_{-1/2}$	$u_3 v^8_{1/2}$
$\psi^7_{-1/2}$	1	0
$\psi^7_{1/2}$	0	1

	$u_3 v^9_{-1/2}$	$u_3 v^9_{1/2}$
$\psi^9_{-1/2}$	0	1
$\psi^9_{1/2}$	–1	0

Table 67. Coupling Coefficients for the Groups D_6, C_{6v}, and D_{3h} (Cont.)

	$u_4 v_{-1}^{5}$	$u_4 v_{1}^{5}$
ψ_{-1}^{6}	i	0
ψ_{1}^{6}	0	–i

	$u_4 v_{-1}^{6}$	$u_4 v_{1}^{6}$
ψ_{-1}^{5}	i	0
ψ_{1}^{5}	0	–i

	$u_4 v_{-1/2}^{7}$	$u_4 v_{1/2}^{7}$
$\psi_{-1/2}^{8}$	i	0
$\psi_{1/2}^{8}$	0	–i

	$u_4 v_{-1/2}^{8}$	$u_4 v_{1/2}^{8}$
$\psi_{-1/2}^{7}$	i	0
$\psi_{1/2}^{7}$	0	–i

	$u_4 v_{-1/2}^{9}$	$u_4 v_{1/2}^{9}$
$\psi_{-1/2}^{9}$	0	i
$\psi_{1/2}^{9}$	i	0

	$u_{-1}^{5} v_{-1}^{5}$	$u_{-1}^{5} v_{1}^{5}$	$u_{1}^{5} v_{-1}^{5}$	$u_{1}^{5} v_{1}^{5}$
ψ_1	0	$1/\sqrt{2}$	$1/\sqrt{2}$	0
ψ_2	0	$i/\sqrt{2}$	$-i/\sqrt{2}$	0
ψ_{-1}^{6}	0	0	0	1
ψ_{1}^{6}	1	0	0	0

	$u_{-1}^{5} v_{-1}^{6}$	$u_{-1}^{5} v_{1}^{6}$	$u_{1}^{5} v_{-1}^{6}$	$u_{1}^{5} v_{1}^{6}$
ψ_3	0	$1/\sqrt{2}$	$1/\sqrt{2}$	0
ψ_4	0	$i/\sqrt{2}$	$-i/\sqrt{2}$	0
ψ_{-1}^{5}	0	0	0	1
ψ_{1}^{5}	1	0	0	0

	$u_{-1}^{5} v_{-1/2}^{7}$	$u_{-1}^{5} v_{1/2}^{7}$	$u_{1}^{5} v_{-1/2}^{7}$	$u_{1}^{5} v_{1/2}^{7}$
$\psi_{-1/2}^{7}$	0	1	0	0
$\psi_{1/2}^{7}$	0	0	–1	0
$\psi_{-1/2}^{9}$	1	0	0	0
ψ^{9}	0	0	0	1

	$u_{-1}^{5} v_{-1/2}^{8}$	$u_{-1}^{5} v_{1/2}^{8}$	$u_{1}^{5} v_{-1/2}^{8}$	$u_{1}^{5} v_{1/2}^{8}$
$\psi_{-1/2}^{8}$	0	1	0	0
$\psi_{1/2}^{8}$	0	0	–1	0
$\psi_{-1/2}^{9}$	0	0	0	1
$\psi_{1/2}^{9}$	–1	0	0	0

	$u_{-1}^{5} v_{-1/2}^{9}$	$u_{-1}^{5} v_{1/2}^{9}$	$u_{1}^{5} v_{-1/2}^{9}$	$u_{1}^{5} v_{1/2}^{9}$
$\psi_{-1/2}^{7}$	0	0	1	0
$\psi_{1/2}^{7}$	0	1	0	0
$\psi_{-1/2}^{8}$	0	0	0	1
$\psi_{1/2}^{8}$	–1	0	0	0

Table 67. Coupling Coefficients for the Groups D_6, C_{6v}, and D_{3h} (Cont.)

	$\overset{6}{u}_{-1}\overset{6}{v}_{-1}$	$\overset{6}{u}_{-1}\overset{6}{v}_{1}$	$\overset{6}{u}_{1}\overset{6}{v}_{-1}$	$\overset{6}{u}_{1}\overset{6}{v}_{1}$
ψ_1	0	$1/\sqrt{2}$	$1/\sqrt{2}$	0
ψ_2	0	$i/\sqrt{2}$	$-i/\sqrt{2}$	0
$\overset{6}{\psi}_{-1}$	0	0	0	1
$\overset{6}{\psi}_{1}$	1	0	0	0

	$\overset{6}{u}_{-1}\overset{7}{v}_{-1/2}$	$\overset{6}{u}_{-1}\overset{7}{v}_{1/2}$	$\overset{6}{u}_{1}\overset{7}{v}_{-1/2}$	$\overset{6}{u}_{1}\overset{7}{v}_{1/2}$
$\overset{8}{\psi}_{-1/2}$	0	1	0	0
$\overset{8}{\psi}_{1/2}$	0	0	−1	0
$\overset{9}{\psi}_{-1/2}$	0	0	0	1
$\overset{9}{\psi}_{1/2}$	−1	0	0	0

	$\overset{6}{u}_{-1}\overset{8}{v}_{-1/2}$	$\overset{6}{u}_{-1}\overset{8}{v}_{1/2}$	$\overset{6}{u}_{1}\overset{8}{v}_{-1/2}$	$\overset{6}{u}_{1}\overset{8}{v}_{1/2}$
$\overset{7}{\psi}_{-1/2}$	0	1	0	0
$\overset{7}{\psi}_{1/2}$	0	0	−1	0
$\overset{9}{\psi}_{-1/2}$	1	0	0	0
$\overset{9}{\psi}_{1/2}$	0	0	0	1

	$\overset{6}{u}_{-1}\overset{9}{v}_{-1/2}$	$\overset{6}{u}_{-1}\overset{9}{v}_{1/2}$	$\overset{6}{u}_{1}\overset{9}{v}_{-1/2}$	$\overset{6}{u}_{1}\overset{9}{v}_{1/2}$
$\overset{7}{\psi}_{-1/2}$	0	0	0	1
$\overset{7}{\psi}_{1/2}$	−1	0	0	0
$\overset{8}{\psi}_{-1/2}$	0	0	1	0
$\overset{8}{\psi}_{1/2}$	0	1	0	0

	$\overset{7}{u}_{-1/2}\overset{7}{v}_{-1/2}$	$\overset{7}{u}_{-1/2}\overset{7}{v}_{1/2}$	$\overset{7}{u}_{1/2}\overset{7}{v}_{-1/2}$	$\overset{7}{u}_{1/2}\overset{7}{v}_{1/2}$
ψ_1	0	$1/\sqrt{2}$	$-1/\sqrt{2}$	0
ψ_2	0	$i/\sqrt{2}$	$i/\sqrt{2}$	0
$\overset{5}{\psi}_{-1}$	1	0	0	0
$\overset{5}{\psi}_{1}$	0	0	0	1

Table 67. Coupling Coefficients for the Groups D_6, C_{6v}, and D_{3h} (Cont.)

	$\overset{7}{u}_{-1/2}\overset{8}{v}_{-1/2}$	$\overset{7}{u}_{-1/2}\overset{8}{v}_{1/2}$	$\overset{7}{u}_{1/2}\overset{8}{v}_{-1/2}$	$\overset{7}{u}_{1/2}\overset{8}{v}_{1/2}$
ψ_3	0	$1/\sqrt{2}$	$-1/\sqrt{2}$	0
ψ_4	0	$i/\sqrt{2}$	$-i/\sqrt{2}$	0
$\overset{6}{\psi}_{-1}$	1	0	0	0
$\overset{6}{\psi}_{1}$	0	0	0	1

	$\overset{7}{u}_{-1/2}\overset{9}{v}_{-1/2}$	$\overset{7}{u}_{-1/2}\overset{9}{v}_{1/2}$	$\overset{7}{u}_{1/2}\overset{9}{v}_{-1/2}$	$\overset{7}{u}_{1/2}\overset{9}{v}_{1/2}$
$\overset{5}{\psi}_{-1}$	0	0	1	0
$\overset{5}{\psi}_{1}$	0	-1	0	0
$\overset{6}{\psi}_{-1}$	0	0	0	1
$\overset{6}{\psi}_{1}$	1	0	0	0

	$\overset{8}{u}_{-1/2}\overset{8}{v}_{-1/2}$	$\overset{8}{u}_{-1/2}\overset{8}{v}_{1/2}$	$\overset{8}{u}_{1/2}\overset{8}{v}_{-1/2}$	$\overset{8}{u}_{1/2}\overset{8}{v}_{1/2}$
ψ_1	0	$1/\sqrt{2}$	$-1/\sqrt{2}$	0
ψ_2	0	$i/\sqrt{2}$	$i/\sqrt{2}$	0
$\overset{5}{\psi}_{-1}$	1	0	0	0
$\overset{5}{\psi}_{1}$	0	0	0	1

	$\overset{8}{u}_{-1/2}\overset{9}{v}_{-1/2}$	$\overset{8}{u}_{-1/2}\overset{9}{v}_{1/2}$	$\overset{8}{u}_{1/2}\overset{9}{v}_{-1/2}$	$\overset{8}{u}_{1/2}\overset{9}{v}_{1/2}$
$\overset{5}{\psi}_{-1}$	0	0	0	1
$\overset{5}{\psi}_{1}$	1	0	0	0
$\overset{6}{\psi}_{-1}$	0	0	1	0
$\overset{6}{\psi}_{1}$	0	-1	0	0

	$\overset{9}{u}_{-1/2}\overset{9}{v}_{-1/2}$	$\overset{9}{u}_{-1/2}\overset{9}{v}_{1/2}$	$\overset{9}{u}_{1/2}\overset{9}{v}_{-1/2}$	$\overset{9}{u}_{1/2}\overset{9}{v}_{1/2}$
ψ_1	0	$1/\sqrt{2}$	$-1/\sqrt{2}$	0
ψ_2	0	$i/\sqrt{2}$	$i/\sqrt{2}$	0
ψ_3	$1/\sqrt{2}$	0	0	$1/\sqrt{2}$
ψ_4	$i/\sqrt{2}$	0	0	$-i/\sqrt{2}$

Table 68. Compatibility Table

D_6	C_{6v}: E(z)	D_{3h}	Γ_1	Γ_2	Γ_3	Γ_4	Γ_5	Γ_6	Γ_7	Γ_8	Γ_9
C_6 :H(z):E(z)	C_6:H(z)	C_{3h}:H(z)	Γ_1	Γ_1	Γ_4	Γ_4	$\Gamma_5 + \Gamma_6$	$\Gamma_2 + \Gamma_3$	$\Gamma_7 + \Gamma_8$	$\Gamma_9 + \Gamma_{10}$	$\Gamma_{11} + \Gamma_{12}$
D_3	C_{3v}	D_3	Γ_1	Γ_2	Γ_1	Γ_2	Γ_3	Γ_3	Γ_4	Γ_4	$\Gamma_5 + \Gamma_6$
		C_{3v}:E(z)	Γ_1	Γ_2	Γ_2	Γ_1	Γ_3	Γ_3	Γ_4	Γ_4	$\Gamma_5 + \Gamma_6$
D_z	C_{2v}		Γ_1	Γ_3	Γ_2	Γ_4	$\Gamma_2 + \Gamma_4$	$\Gamma_1 + \Gamma_3$	Γ_5	Γ_5	Γ_5
		C_{2v}:E(y)	Γ_1	Γ_2	Γ_3	Γ_4	$\Gamma_3 + \Gamma_4$	$\Gamma_1 + \Gamma_2$	Γ_5	Γ_5	Γ_5
C_2:H(y):E(y)	C_s :E(x):H(y)	C_2 :H(y)	Γ_1	Γ_2	Γ_1	Γ_2	$\Gamma_1 + \Gamma_2$	$\Gamma_1 + \Gamma_2$	$\Gamma_3 + \Gamma_4$	$\Gamma_3 + \Gamma_4$	$\Gamma_3 + \Gamma_4$
C_2 :H(x):E(x)	C_s :H(x):E(y)	C_s :H(x)	Γ_1	Γ_2	Γ_2	Γ_1	$\Gamma_1 + \Gamma_2$	$\Gamma_1 + \Gamma_2$	$\Gamma_3 + \Gamma_4$	$\Gamma_3 + \Gamma_4$	$\Gamma_3 + \Gamma_4$

Table 69. Full Rotation Group Compatibility Table for the Group D_6

$D_0^\pm$	Γ_1
$D_1^\pm$	$\Gamma_2 + \Gamma_5$
$D_2^\pm$	$\Gamma_1 + \Gamma_5 + \Gamma_6$
$D_3^\pm$	$\Gamma_2 + \Gamma_3 + \Gamma_4 + \Gamma_5 + \Gamma_6$
$D_4^\pm$	$\Gamma_1 + \Gamma_3 + \Gamma_4 + \Gamma_5 + 2\Gamma_6$
$D_5^\pm$	$\Gamma_2 + \Gamma_3 + \Gamma_4 + 2\Gamma_5 + 2\Gamma_6$
$D_6^\pm$	$2\Gamma_1 + \Gamma_2 + \Gamma_3 + \Gamma_4 + 2\Gamma_5 + 2\Gamma_6$
$D_{1/2}^\pm$	Γ_7
$D_{3/2}^\pm$	$\Gamma_7 + \Gamma_9$
$D_{5/2}^\pm$	$\Gamma_7 + \Gamma_8 + \Gamma_9$
$D_{7/2}^\pm$	$\Gamma_7 + 2\Gamma_8 + \Gamma_9$
$D_{9/2}^\pm$	$\Gamma_7 + 2\Gamma_8 + 2\Gamma_9$
$D_{11/2}^\pm$	$2\Gamma_7 + 2\Gamma_8 + 2\Gamma_9$
$D_{13/2}^\pm$	$3\Gamma_7 + 2\Gamma_8 + 2\Gamma_9$

Table 70. Full Rotation Group Compatibility Table for the Group C_{6v}

D_0^+	Γ_1	D_0^-	Γ_2
D_1^+	$\Gamma_2 + \Gamma_5$	D_1^-	$\Gamma_1 + \Gamma_5$
D_2^+	$\Gamma_1 + \Gamma_5 + \Gamma_6$	D_2^-	$\Gamma_2 + \Gamma_5 + \Gamma_6$
D_3^+	$\Gamma_2 + \Gamma_3 + \Gamma_4 + \Gamma_5 + \Gamma_6$	D_3^-	$\Gamma_1 + \Gamma_3 + \Gamma_4 + \Gamma_5 + \Gamma_6$
D_4^+	$\Gamma_1 + \Gamma_3 + \Gamma_4 + \Gamma_5 + 2\Gamma_6$	D_4^-	$\Gamma_2 + \Gamma_3 + \Gamma_4 + \Gamma_5 + 2\Gamma_6$
D_5^+	$\Gamma_2 + \Gamma_3 + \Gamma_4 + 2\Gamma_5 + 2\Gamma_6$	D_5^-	$\Gamma_1 + \Gamma_3 + \Gamma_4 + 2\Gamma_5 + 2\Gamma_6$
D_6^+	$2\Gamma_1 + \Gamma_2 + \Gamma_3 + \Gamma_4 + 2\Gamma_5 + 2\Gamma_6$	D_6^-	$\Gamma_1 + 2\Gamma_2 + \Gamma_3 + \Gamma_4 + 2\Gamma_5 + 2\Gamma_6$
$D_{1/2}^+$	Γ_7	$D_{1/2}^-$	Γ_7
$D_{3/2}^\pm$	$\Gamma_7 + \Gamma_9$	$D_{3/2}^-$	$\Gamma_7 + \Gamma_9$
$D_{5/2}^+$	$\Gamma_7 + \Gamma_8 + \Gamma_9$	$D_{5/2}^-$	$\Gamma_7 + \Gamma_8 + \Gamma_9$
$D_{7/2}^+$	$\Gamma_7 + 2\Gamma_8 + \Gamma_9$	$D_{7/2}^-$	$\Gamma_7 + 2\Gamma_8 + \Gamma_9$
$D_{9/2}^+$	$\Gamma_7 + 2\Gamma_8 + 2\Gamma_9$	$D_{9/2}^-$	$\Gamma_7 + 2\Gamma_8 + 2\Gamma_9$
$D_{11/2}^+$	$2\Gamma_7 + 2\Gamma_8 + 2\Gamma_9$	$D_{11/2}^-$	$2\Gamma_7 + 2\Gamma_8 + 2\Gamma_9$
$D_{13/2}^+$	$3\Gamma_7 + 2\Gamma_8 + 2\Gamma_9$	$D_{13/2}^-$	$3\Gamma_7 + 2\Gamma_8 + 2\Gamma_9$

Table 71. Full Rotation Group Compatibility Table for the Group D_{3h}

D_0^+	Γ_1	D_0^-	Γ_3
D_1^+	$\Gamma_2 + \Gamma_5$	D_1^-	$\Gamma_4 + \Gamma_6$
D_2^+	$\Gamma_1 + \Gamma_5 + \Gamma_6$	D_2^-	$\Gamma_3 + \Gamma_5 + \Gamma_6$
D_3^+	$\Gamma_2 + \Gamma_3 + \Gamma_4 + \Gamma_5 + \Gamma_6$	D_3^-	$\Gamma_1 + \Gamma_2 + \Gamma_4 + \Gamma_5 + \Gamma_6$
D_4^+	$\Gamma_1 + \Gamma_3 + \Gamma_4 + \Gamma_5 + 2\Gamma_6$	D_4^-	$\Gamma_1 + \Gamma_2 + \Gamma_3 + 2\Gamma_5 + \Gamma_6$
D_5^+	$\Gamma_2 + \Gamma_3 + \Gamma_4 + 2\Gamma_5 + 2\Gamma_6$	D_5^-	$\Gamma_1 + \Gamma_2 + \Gamma_4 + 2\Gamma_5 + 2\Gamma_6$
D_6^+	$2\Gamma_1 + \Gamma_2 + \Gamma_3 + \Gamma_4 + 2\Gamma_5 + 2\Gamma_6$	D_6^-	$\Gamma_1 + \Gamma_2 + 2\Gamma_3 + \Gamma_4 + 2\Gamma_5 + 2\Gamma_6$
$D_{1/2}^+$	Γ_7	$D_{1/2}^-$	Γ_8
$D_{3/2}^+$	$\Gamma_7 + \Gamma_9$	$D_{3/2}^-$	$\Gamma_8 + \Gamma_9$
$D_{5/2}^+$	$\Gamma_7 + \Gamma_8 + \Gamma_9$	$D_{5/2}^-$	$\Gamma_7 + \Gamma_8 + \Gamma_9$
$D_{7/2}^+$	$\Gamma_7 + 2\Gamma_8 + \Gamma_9$	$D_{7/2}^-$	$2\Gamma_7 + \Gamma_8 + \Gamma_9$
$D_{9/2}^+$	$\Gamma_7 + 2\Gamma_8 + 2\Gamma_9$	$D_{9/2}^-$	$2\Gamma_7 + \Gamma_8 + 2\Gamma_9$
$D_{11/2}^+$	$2\Gamma_7 + 2\Gamma_8 + 2\Gamma_9$	$D_{11/2}^-$	$2\Gamma_7 + 2\Gamma_8 + 2\Gamma_9$
$D_{13/2}$	$3\Gamma_7 + 2\Gamma_8 + 2\Gamma_9$	$D_{13/2}$	$2\Gamma_7 + 3\Gamma_8 + 2\Gamma_9$

(27) D_{6h}. This group can be considered as the direct product of D_6 and C_i ($D_{6h} = D_6 \times C_i$). It may also be considered as the direct product of D_6 and C_s, of C_{6v} and C_i, of C_{6v} and C_s, of D_{3h} and C_i, and of D_{3h} and C_2. It contains all the operations of these groups. If in Figure 3 we take the sixfold axis as z the reflection σ_h occurs in the xy plane, the six twofold axes are y, y′, y″, ($C_{2'}$) and x, x′, x″ ($C_{2''}$), the six reflection planes are xz, x′z, x″z (σ_d) and yz, y′z, y″z (σ_v). In addition to these operations we have $S_6 = C_3^{-1} \times I$, $S_6^{-1} = C_3 \times I$; $S_3 = C_6^{-1} \times I$ and $S_3^{-1} = C_6 \times I$.

Table 72. Character Table

D_{6h}	E	$\bar{E}$	C_2 $\bar{C}_2$	$2C_3$	$2\bar{C}_3$	$2C_6$	$2\bar{C}_6$	$3C_2'$ $3\bar{C}_2'$	$3C_2''$ $3\bar{C}_2''$
Γ_1^+	1	1	1	1	1	1	1	1	1
Γ_2^+	1	1	1	1	1	1	1	−1	−1
Γ_3^+	1	1	−1	1	1	−1	−1	1	−1
Γ_4^+	1	1	−1	1	1	−1	−1	−1	1
Γ_5^+	2	2	−2	−1	−1	1	1	0	0
Γ_6^+	2	2	2	−1	−1	−1	−1	0	0
Γ_1^-	1	1	1	1	1	1	1	1	1
Γ_2^-	1	1	1	1	1	1	1	−1	−1
Γ_3^-	1	1	−1	1	1	−1	−1	1	−1
Γ_4^-	1	1	−1	1	1	−1	−1	−1	1
Γ_5^-	2	2	−2	−1	−1	1	1	0	0
Γ_6^-	2	2	2	−1	−1	−1	−1	0	0
Γ_7^+	2	−2	0	1	−1	$\sqrt{3}$	$-\sqrt{3}$	0	0
Γ_8^+	2	−2	0	1	−1	$-\sqrt{3}$	$\sqrt{3}$	0	0
Γ_9^+	2	−2	0	−2	2	0	0	0	0
Γ_7^-	2	−2	0	1	−1	$\sqrt{3}$	$-\sqrt{3}$	0	0
Γ_8^-	2	−2	0	1	−1	$-\sqrt{3}$	$\sqrt{3}$	0	0
Γ_9^-	2	−2	0	−2	2	0	0	0	0

and Basis Functions for the Group D_{6h}

D_{6h}	I	$\bar{I}$	σ_h σ_h	$2S_6$	$2\bar{S}_6$	$2S_3$	$2\bar{S}_3$	$3\sigma_d$ $3\bar{\sigma}_d$	$3\sigma_v$ $3\bar{\sigma}_v$	Time Inv.	Bases
Γ_1^+	1	1	1	1	1	1	1	1	1	a	R
Γ_2^+	1	1	1	1	1	1	1	−1	−1	a	S_z
Γ_3^+	1	1	−1	1	1	−1	−1	1	−1	a	$z(x^3 - 3xy^2)$
Γ_4^+	1	1	−1	1	1	−1	−1	−1	1	a	$z(y^3 - 3yx^2)$
Γ_5^+	2	2	−2	−1	−1	1	1	0	0	a	$(S_x - iSy)$, $-(S_x + iSy)$
Γ_6^+	2	2	2	−1	−1	−1	−1	0	0	a	$\Gamma_5^+ \times \Gamma_3^+$
Γ_1^-	−1	−1	−1	−1	−1	−1	−1	−1	−1	a	zS_z
Γ_2^-	−1	−1	−1	−1	−1	−1	−1	1	1	a	z
Γ_3^-	−1	−1	1	−1	−1	1	1	−1	1	a	$y^3 - 3yx^2$
Γ_4^-	−1	−1	1	−1	−1	1	1	1	−1	a	$x^3 - 3xy^2$
Γ_5^-	−2	−2	2	1	1	−1	−1	0	0	a	$(x - iy)$,$-(x + iy)$
Γ_6^-	−2	−2	−2	1	1	1	1	0	0	a	$\Gamma_5^+ \times \Gamma_3^-$
Γ_7^+	2	−2	0	1	−1	$\sqrt{3}$	$-\sqrt{3}$	0	0	c	$\phi(1/2,-1/2)$, $\phi(1/2, 1/2)$
Γ_8^+	2	−2	0	1	−1	$-\sqrt{3}$	$\sqrt{3}$	0	0	c	$\Gamma_7^+ \times \Gamma_3^+$
Γ_9^+	2	−2	0	−2	2	0	0	0	0	c	$\phi(3/2,-3/2)$, $\phi(3/2, 3/2)$
Γ_7^-	−2	2	0	−1	1	$-\sqrt{3}$	$\sqrt{3}$	0	0	c	$\Gamma_7^+ \times \Gamma_1^-$
Γ_8^-	−2	2	0	−1	1	$\sqrt{3}$	$-\sqrt{3}$	0	0	c	$\Gamma_7^+ \times \Gamma_3^-$
Γ_9^-	−2	2	0	2	−2	0	0	0	0	c	$\Gamma_9^+ \times \Gamma_1^-$

Table 73. Compatibility Table

D_{6h}	Γ_1^+	Γ_2^+	Γ_3^+	Γ_4^+	Γ_5^+	Γ_6^+	Γ_1^-	Γ_2^-	Γ_3^-	Γ_4^-	Γ_5^-	Γ_6^-	Γ_7^+	Γ_8^+	Γ_9^+	Γ_4^-	Γ_4^-	Γ_9^-
D_6	Γ_1	Γ_2	Γ_3	Γ_4	Γ_5	Γ_6	Γ_1	Γ_2	Γ_3	Γ_4	Γ_5	Γ_6	Γ_7	Γ_8	Γ_9	Γ_7	Γ_8	Γ_9
C_{6h}:H(z)	Γ_1^+	Γ_1^+	Γ_4^+	Γ_4^+	$\Gamma_5^+ + \Gamma_6^+$	$\Gamma_2^+ + \Gamma_3^+$	Γ_1^-	Γ_1^-	Γ_4^-	Γ_4^-	$\Gamma_5^- + \Gamma_6^-$	$\Gamma_2^- + \Gamma_3^-$	$\Gamma_7^+ + \Gamma_8^+$	$\Gamma_9^+ + \Gamma_{10}^+$	$\Gamma_{11}^+ + \Gamma_{12}^+$	$\Gamma_7^- + \Gamma_8^-$	$\Gamma_9^- + \Gamma_{10}^-$	$\Gamma_{11}^- + \Gamma_{12}^-$
C_{6v}:E(z)	Γ_1	Γ_2	Γ_3	Γ_4	Γ_5	Γ_6	Γ_2	Γ_1	Γ_4	Γ_3	Γ_5	Γ_6	Γ_7	Γ_8	Γ_9	Γ_7	Γ_8	Γ_9
D_{3d}	Γ_1^+	Γ_2^+	Γ_1^+	Γ_2^+	Γ_3^+	Γ_3^+	Γ_1^-	Γ_2^-	Γ_1^-	Γ_2^-	Γ_3^-	Γ_3^-	Γ_4^+	Γ_4^+	$\Gamma_5^+ + \Gamma_6^+$	Γ_4^-	Γ_4^-	$\Gamma_5^- + \Gamma_6^-$
D_{3h}	Γ_1	Γ_2	Γ_4	Γ_3	Γ_5	Γ_6	Γ_3	Γ_4	Γ_2	Γ_1	Γ_6	Γ_5	Γ_7	Γ_8	Γ_9	Γ_8	Γ_7	Γ_9
D_{2h}	Γ_1^+	Γ_3^+	Γ_2^+	Γ_4^+	$\Gamma_2^+ + \Gamma_4^+$	$\Gamma_1^+ + \Gamma_3^+$	Γ_1^-	Γ_3^-	Γ_2^-	Γ_4^-	$\Gamma_2^- + \Gamma_4^-$	$\Gamma_1^- + \Gamma_3^-$	Γ_5^+	Γ_5^+	Γ_5^+	Γ_5^-	Γ_5^-	Γ_5^-
C_{2h}:H(y)	Γ_1^+	Γ_2^+	Γ_1^+	Γ_2^+	$\Gamma_1^+ + \Gamma_2^+$	$\Gamma_1^+ + \Gamma_2^+$	Γ_1^-	Γ_2^-	Γ_1^-	Γ_2^-	$\Gamma_1^- + \Gamma_2^-$	$\Gamma_1^- + \Gamma_2^-$	$\Gamma_3^+ + \Gamma_4^+$	$\Gamma_3^+ + \Gamma_4^+$	$\Gamma_3^+ + \Gamma_4^+$	$\Gamma_3^- + \Gamma_4^-$	$\Gamma_3^- + \Gamma_4^-$	$\Gamma_3^- + \Gamma_4^-$
C_{2v}:E(y)	Γ_1	Γ_2	Γ_3	Γ_4	$\Gamma_3 + \Gamma_4$	$\Gamma_1 + \Gamma_2$	Γ_3	Γ_4	Γ_1	Γ_2	$\Gamma_1 + \Gamma_2$	$\Gamma_3 + \Gamma_4$	Γ_5	Γ_5	Γ_5	Γ_5	Γ_5	Γ_5

(28) T. This group consists of all proper rotations which send a regular tetrahedron into itself. There are four threefold axes and three twofold axes. This is best illustrated by a diagram. If we call the center of the cube in Figure 4, O, and consider the tetrahedron 1234 whose vertices coincide with four of the vertices of the cube, the four threefold axes are O1, O2, O3, and O4 (C_3 and C_3^{-1}). The three twofold axes would then coincide with the x, y, z axes drawn from the origin O to the midpoints of the faces of the cube.

Table 74. Character Table and Basis Functions for the Group T

T	E	$\bar{E}$	$3C_2$ $3\bar{C}_2$	$4C_3$	$4\bar{C}_3$	$4C_3^{-1}$	$4\bar{C}_3^{-1}$	Time Inv.	Bases
Γ_1	1	1	1	1	1	1	1	a	R or xyz
Γ_2	1	1	1	ω	ω	ω^2	ω^2	b	$1/\sqrt{2}\ (u+iv)$
Γ_3	1	1	1	ω^2	ω^2	ω	ω	b	$1/\sqrt{2}\ (u-iv)$
Γ_4	3	3	-1	0	0	0	0	a	S_x, S_y, S_z
Γ_5	2	-2	0	1	-1	1	-1	c	$\phi(1/2,-1/2)$, $\phi(1/2,1/2)$
Γ_6	2	-2	0	ω	$-\omega$	ω^2	$-\omega^2$	b	$\psi^6_{-1/2}$, $\psi^6_{1/2}$
Γ_7	2	-2	0	ω^2	$-\omega^2$	ω	$-\omega$	b	Γ_6^*

$\omega = \exp\,(2\pi i/3)$

$$u = 3z^2 - r^2 \qquad v = \sqrt{3}\,(x^2 - y^2)$$

$$\psi^6_{-1/2} = 1/\sqrt{2}\left(|-\tfrac{1}{2}> - i\,|\tfrac{3}{2}>\right) \qquad \psi^6_{1/2} = -1/\sqrt{2}\left(|\tfrac{1}{2}> - i\,|-\tfrac{3}{2}>\right)$$

$$\psi^7_{-1/2} = 1/\sqrt{2}\left(|\tfrac{1}{2}> + i\,|-\tfrac{3}{2}>\right) \qquad \psi^7_{1/2} = 1/\sqrt{2}\left(|-\tfrac{1}{2}> + i\,|\tfrac{3}{2}>\right)$$

Table 75. Multiplication Table for the Group T

Γ_1	Γ_2	Γ_3	Γ_4	Γ_5	Γ_6	Γ_7	
Γ_1	Γ_2	Γ_3	Γ_4	Γ_5	Γ_6	Γ_7	Γ_1
	Γ_3	Γ_1	Γ_4	Γ_6	Γ_7	Γ_5	Γ_2
		Γ_2	Γ_4	Γ_7	Γ_5	Γ_6	Γ_3
			$\Gamma_1 + \Gamma_2 + \Gamma_3 + 2\Gamma_4$	$\Gamma_5 + \Gamma_6 + \Gamma_7$	$\Gamma_5 + \Gamma_6 + \Gamma_7$	$\Gamma_5 + \Gamma_6 + \Gamma_7$	Γ_4
				$\Gamma_1 + \Gamma_4$	$\Gamma_2 + \Gamma_4$	$\Gamma_3 + \Gamma_4$	Γ_5
					$\Gamma_3 + \Gamma_4$	$\Gamma_1 + \Gamma_4$	Γ_6
						$\Gamma_2 + \Gamma_4$	Γ_7

Table 76. Coupling Coefficients for the Group T

$u_3 = \mathbf{K}u_2; \quad u_2 = \mathbf{K}u_3$

$\overset{7}{u}_{-1/2} = \mathbf{K}\overset{6}{u}_{-1/2}; \quad \overset{6}{u}_{-1/2} = -\mathbf{K}\overset{7}{u}_{-1/2}$

$\overset{7}{u}_{1/2} = \mathbf{K}\overset{6}{u}_{1/2}; \quad \overset{6}{u}_{1/2} = -\mathbf{K}\overset{7}{u}_{1/2}$

$u_2 v_2 = \psi_3 \qquad u_3 v_3 = \psi_2$

	$u_2 v_3$	$u_3 v_2$
ψ_1	$1/\sqrt{2}$	$1/\sqrt{2}$
ψ_1	$i/\sqrt{2}$	$-i/\sqrt{2}$

	$u_2 \overset{4}{v}_x$	$u_2 \overset{4}{v}_y$	$u_2 \overset{4}{v}_z$	$u_3 \overset{4}{v}_x$	$u_3 \overset{4}{v}_y$	$u_3 \overset{4}{v}_z$
$\overset{4}{\psi}_x$	$1/\sqrt{2}$	0	0	$1/\sqrt{2}$	0	0
$\overset{4}{\psi}_y$	0	$\omega/\sqrt{2}$	0	0	$\omega^2/\sqrt{2}$	0
$\overset{4}{\psi}_z$	0	0	$\omega^2/\sqrt{2}$	0	0	$\omega/\sqrt{2}$
$\overset{4}{\psi}_x$	$i/\sqrt{2}$	0	0	$-i/\sqrt{2}$	0	0
$\overset{4}{\psi}_y$	0	$i\omega/\sqrt{2}$	0	0	$-i\omega^2/\sqrt{2}$	0
$\overset{4}{\psi}_z$	0	0	$i\omega^2/\sqrt{2}$	0	0	$-i\omega/\sqrt{2}$

	$u_2 \overset{5}{v}_{-1/2}$	$u_2 \overset{5}{v}_{1/2}$
$\overset{6}{\psi}_{-1/2}$	1	0
$\overset{6}{\psi}_{1/2}$	0	1

	$u_3 \overset{5}{v}_{-1/2}$	$u_3 \overset{5}{v}_{1/2}$
$\overset{7}{\psi}_{-1/2}$	0	-1
$\overset{7}{\psi}_{1/2}$	1	0

	$u_2 \overset{6}{v}_{-1/2}$	$u_2 \overset{6}{v}_{1/2}$
$\overset{7}{\psi}_{-1/2}$	0	-1
$\overset{7}{\psi}_{1/2}$	1	0

	$u_3 \overset{7}{v}_{-1/2}$	$u_3 \overset{7}{v}_{1/2}$
$\overset{6}{\psi}_{-1/2}$	0	1
$\overset{6}{\psi}_{1/2}$	-1	0

$\omega = \exp(2\pi i/3)$

Table 76. Coupling Coefficients for the Group T (Cont.)

	$u_2 v^7_{-1/2}$	$u_2 v^7_{1/2}$	$u_3 v^7_{-1/2}$	$u_3 v^7_{1/2}$
$\psi^5_{-1/2}$	0	$1/\sqrt{2}$	$1/\sqrt{2}$	0
$\psi^5_{1/2}$	$-1/\sqrt{2}$	0	0	$1/\sqrt{2}$
$\psi^5_{-1/2}$	0	$i/\sqrt{2}$	$-i/\sqrt{2}$	0
$\psi^5_{1/2}$	$-i/\sqrt{2}$	0	0	$-i/\sqrt{2}$

	$u^4_x v^4_x$	$u^4_x v^4_y$	$u^4_x v^4_z$	$u^4_y v^4_x$	$u^4_y v^4_y$	$u^4_y v^4_z$	$u^4_z v^4_x$	$u^4_z v^4_y$	$u^4_z v^4_z$
ψ_1	$1/\sqrt{3}$	0	0	0	$1/\sqrt{3}$	0	0	0	$1/\sqrt{3}$
ψ_2	$\omega^2/\sqrt{3}$	0	0	0	$\omega/\sqrt{3}$	0	0	0	$1/\sqrt{3}$
ψ_3	$\omega/\sqrt{3}$	0	0	0	$\omega^2/\sqrt{3}$	0	0	0	$1/\sqrt{3}$
ψ^4_x	0	0	0	0	0	$1/\sqrt{2}$	0	$1/\sqrt{2}$	0
ψ^4_y	0	0	$1/\sqrt{2}$	0	0	0	$1/\sqrt{2}$	0	0
ψ^4_z	0	$1/\sqrt{2}$	0	$1/\sqrt{2}$	0	0	0	0	0
ϕ^4_x	0	0	0	0	0	$1/\sqrt{2}$	0	$-1/\sqrt{2}$	0
ϕ^4_y	0	0	$-1/\sqrt{2}$	0	0	0	$1/\sqrt{2}$	0	0
ϕ^4_z	0	$1/\sqrt{2}$	0	$-1/\sqrt{2}$	0	0	0	0	0

	$u^5_x v^6_{-1/2}$	$u^5_x v^6_{1/2}$	$u^5_y v^6_{-1/2}$	$u^5_y v^6_{1/2}$	$u^5_z v^6_{-1/2}$	$u^5_z v^6_{1/2}$
$\psi^5_{-1/2}$	0	$-i/\sqrt{3}$	0	$-1/\sqrt{3}$	$i/\sqrt{3}$	0
$\psi^5_{1/2}$	$-i/\sqrt{3}$	0	$1/\sqrt{3}$	0	0	$-i/\sqrt{3}$
$\psi^6_{-1/2}$	0	$-i\omega^2/\sqrt{3}$	0	$-\omega/\sqrt{3}$	$i/\sqrt{3}$	0
$\psi^6_{1/2}$	$-i\omega^2/\sqrt{3}$	0	$\omega/\sqrt{3}$	0	0	$-i/\sqrt{3}$
$\psi^7_{-1/2}$	$i\omega/\sqrt{3}$	0	$-\omega^2/\sqrt{3}$	0	0	$i/\sqrt{3}$
$\psi^7_{1/2}$	0	$-i\omega/\sqrt{3}$	0	$-\omega^2/\sqrt{3}$	$i/\sqrt{3}$	0

$\omega = \exp(2\pi i/3)$

Table 76. Coupling Coefficients for the Group T (Cont.)

	$u_x^4 v_{-1/2}^6$	$u_x^4 v_{1/2}^6$	$u_x^4 v_{-1/2}^7$	$u_x^4 v_{1/2}^7$	$u_y^4 v_{-1/2}^6$	$u_y^4 v_{1/2}^6$	$u_y^4 v_{-1/2}^7$	$u_y^4 v_{1/2}^7$	$u_z^4 v_{-1/2}^6$	$u_z^4 v_{1/2}^6$	$u_z^4 v_{-1/2}^7$	$u_z^4 v_{1/2}^7$
$\psi_{-1/2}^5$	0	$-i\omega/\sqrt{6}$	$i\omega^2/\sqrt{6}$	0	0	$-\omega^2/\sqrt{6}$	$\omega/\sqrt{6}$	0	$i/\sqrt{6}$	0	0	$i/\sqrt{6}$
$\psi_{1/2}^5$	$-i\omega/\sqrt{6}$	0	0	$-i\omega^2/\sqrt{6}$	$\omega^2/\sqrt{6}$	0	0	$\omega/\sqrt{6}$	0	$-i/\sqrt{6}$	$i/\sqrt{6}$	0
$\psi_{-1/2}^5$	0	$\omega/\sqrt{6}$	$\omega^2/\sqrt{6}$	0	0	$-i\omega^2/\sqrt{6}$	$-i\omega/\sqrt{6}$	0	$-1/\sqrt{6}$	0	0	$1/\sqrt{6}$
$\psi_{1/2}^5$	$\omega/\sqrt{6}$	0	0	$-\omega^2/\sqrt{6}$	$i\omega^2/\sqrt{6}$	0	0	$-i\omega/\sqrt{6}$	0	$1/\sqrt{6}$	$1/\sqrt{6}$	0
$\psi_{-1/2}^6$	0	$-i/\sqrt{3}$	0	0	0	$-1/\sqrt{3}$	0	0	$i/\sqrt{3}$	0	0	0
$\psi_{1/2}^6$	$-i/\sqrt{3}$	0	0	0	$1/\sqrt{3}$	0	0	0	0	$-i/\sqrt{3}$	0	0
$\psi_{-1/2}^7$	0	0	0	$i/\sqrt{3}$	0	0	0	$-1/\sqrt{3}$	0	0	$-i/\sqrt{3}$	0
$\psi_{1/2}^7$	0	0	$i/\sqrt{3}$	0	0	0	$1/\sqrt{3}$	0	0	0	0	$i/\sqrt{3}$
$\psi_{-1/2}^6$	0	0	$i\omega/\sqrt{3}$	0	0	0	$\omega^2/\sqrt{3}$	0	0	0	0	$i/\sqrt{3}$
$\psi_{1/2}^6$	0	0	0	$-i\omega/\sqrt{3}$	0	0	0	$\omega^2/\sqrt{3}$	0	0	$i/\sqrt{3}$	0
$\psi_{-1/2}^7$	$i\omega^2/\sqrt{3}$	0	0	0	$-\omega/\sqrt{3}$	0	0	0	0	$i/\sqrt{3}$	0	0
$\psi_{1/2}^7$	0	$-i\omega^2/\sqrt{3}$	0	0	0	$-\omega/\sqrt{3}$	0	0	$i/\sqrt{3}$	0	0	0

$\omega = \exp(2\pi i/3)$

Table 76. Coupling Coefficients for the Group T (Cont.)

	$u^5_{-1/2}v^5_{-1/2}$	$u^5_{-1/2}v^5_{1/2}$	$u^5_{1/2}v^5_{-1/2}$	$u^5_{1/2}v^5_{1/2}$
ψ_1	0	$-1/\sqrt{2}$	$1/\sqrt{2}$	0
ψ^4_x	$-i/\sqrt{2}$	0	0	$i/\sqrt{2}$
ψ^4_y	$1/\sqrt{2}$	0	0	$1/\sqrt{2}$
ψ^4_z	0	$-i/\sqrt{2}$	$-i/\sqrt{2}$	0

	$u^5_{-1/2}v^6_{-1/2}$	$u^5_{-1/2}v^6_{1/2}$	$u^5_{-1/2}v^7_{-1/2}$	$u^5_{-1/2}v^7_{1/2}$	$u^5_{1/2}v^6_{-1/2}$	$u^5_{1/2}v^6_{1/2}$	$u^5_{1/2}v^7_{-1/2}$	$u^5_{1/2}v^7_{1/2}$
ψ_2	0	$-1/\sqrt{2}$	0	0	$1/\sqrt{2}$	0	0	0
ψ_3	0	0	$1/\sqrt{2}$	0	0	0	0	$1/\sqrt{2}$
ψ^4_x	$-i/2$	0	0	$-i/2$	0	$i/2$	$-i/2$	0
ψ^4_y	$\omega/2$	0	0	$\omega^2/2$	0	$\omega/2$	$-\omega^2/2$	0
ψ^4_z	0	$-i\omega^2/2$	$i\omega/2$	0	$-i\omega^2/2$	0	0	$-i\omega/2$
ψ^4_x	$1/2$	0	0	$-1/2$	0	$-1/2$	$-1/2$	0
ψ^4_y	$i\omega/2$	0	0	$-i\omega^2/2$	0	$i\omega/2$	$i\omega^2/2$	0
ψ^4_z	0	$\omega^2/2$	$\omega/2$	0	$\omega^2/2$	0	0	$-\omega/2$

$\omega = \exp(2\pi i/3)$

Table 76. Coupling Coefficients for the Group T (Cont.)

	$u^6_{-1/2}v^6_{-1/2}$	$u^6_{-1/2}v^6_{1/2}$	$u^6_{-1/2}v^7_{-1/2}$	$u^6_{-1/2}v^7_{1/2}$	$u^6_{1/2}v^6_{-1/2}$	$u^6_{1/2}v^6_{1/2}$	$u^6_{1/2}v^7_{-1/2}$	$u^6_{1/2}v^7_{1/2}$	$u^7_{-1/2}v^6_{-1/2}$	$u^7_{-1/2}v^6_{1/2}$	$u^7_{-1/2}v^7_{-1/2}$	$u^7_{-1/2}v^7_{1/2}$	$u^7_{1/2}v^6_{-1/2}$	$u^7_{1/2}v^6_{1/2}$	$u^7_{1/2}v^7_{-1/2}$	$u^7_{1/2}v^7_{1/2}$
ψ_1	0	0	1/2	0	0	0	0	1/2	−1/2	0	0	0	0	−1/2	0	0
ψ_1	0	0	i/2	0	0	0	0	i/2	i/2	0	0	0	0	i/2	0	0
ψ_2	0	0	0	0	0	0	0	0	0	0	0	$-1/\sqrt{2}$	0	0	$1/\sqrt{2}$	0
ψ_3	0	$-1/\sqrt{2}$	0	0	$1/\sqrt{2}$	0	0	0	0	0	0	0	0	0	0	0
ψ^4_x	−i/2	0	0	0	0	i/2	0	0	0	0	i/2	0	0	0	0	−i/2
ψ^4_y	$\omega^2/2$	0	0	0	0	$\omega^2/2$	0	0	0	0	$\omega/2$	0	0	0	0	$\omega/2$
ψ^4_z	0	$-i\omega/2$	0	0	$-i\omega/2$	0	0	0	0	0	0	$i\omega^2/2$	0	0	$i\omega^2/2$	0
ψ^4_x	1/2	0	0	0	0	−1/2	0	0	0	0	1/2	0	0	0	0	−1/2
ψ^4_y	$i\omega^2/2$	0	0	0	0	$i\omega^2/2$	0	0	0	0	$-i\omega/2$	0	0	0	0	$-i\omega/2$
ψ^4_z	0	$\omega/2$	0	0	$\omega/2$	0	0	0	0	0	0	$\omega^2/2$	0	0	$\omega^2/2$	0
ψ^4_x	0	0	0	−1/2	0	0	−1/2	0	0	1/2	0	0	1/2	0	0	0
ψ^4_y	0	0	0	−i/2	0	0	i/2	0	0	i/2	0	0	−i/2	0	0	0
ψ^4_z	0	0	1/2	0	0	0	0	−1/2	1/2	0	0	0	0	−1/2	0	0
ψ^4_x	0	0	0	−i/2	0	0	−i/2	0	0	−i/2	0	0	−i/2	0	0	0
ψ^4_y	0	0	0	1/2	0	0	−1/2	0	0	1/2	0	0	−1/2	0	0	0
ψ^4_z	0	0	i/2	0	0	0	0	−i/2	−i/2	0	0	0	0	i/2	0	0

$\omega = \exp(2\pi i/3)$

Table 77. Compatibility Table

T	Γ_1	Γ_2	Γ_3	Γ_4	Γ_5	Γ_6	Γ_7
D_2	Γ_1	Γ_1	Γ_1	$\Gamma_2 + \Gamma_3 + \Gamma_4$	Γ_5	Γ_5	Γ_5
C_3 :H(w):E(w)	Γ_1	Γ_2	Γ_3	$\Gamma_1 + \Gamma_2 + \Gamma_3$	$\Gamma_4 + \Gamma_5$	$\Gamma_4 + \Gamma_6$	$\Gamma_5 + \Gamma_6$
C_2 :H(z):E(z)	Γ_1	Γ_1	Γ_1	$\Gamma_1 + 2\Gamma_2$	$\Gamma_3 + \Gamma_4$	$\Gamma_3 + \Gamma_4$	$\Gamma_3 + \Gamma_4$

Table 78. Full Rotation Group Compatibility Table for the Group T

$D_0^{\pm}$	Γ_1
$D_1^{\pm}$	Γ_4
$D_2^{\pm}$	$\Gamma_2 + \Gamma_3 + \Gamma_4$
$D_3^{\pm}$	$\Gamma_1 + 2\Gamma_4$
$D_4^{\pm}$	$\Gamma_1 + \Gamma_2 + \Gamma_3 + 2\Gamma_4$
$D_5^{\pm}$	$\Gamma_2 + \Gamma_3 + 3\Gamma_4$
$D_6^{\pm}$	$2\Gamma_1 + \Gamma_2 + \Gamma_3 + 3\Gamma_4$
$D_{1/2}^{\pm}$	Γ_5
$D_{3/2}^{\pm}$	$\Gamma_6 + \Gamma_7$
$D_{5/2}^{\pm}$	$\Gamma_5 + \Gamma_6 + \Gamma_7$
$D_{7/2}^{\pm}$	$2\Gamma_5 + \Gamma_6 + \Gamma_7$
$D_{9/2}^{\pm}$	$\Gamma_5 + 2\Gamma_6 + 2\Gamma_7$
$D_{11/2}^{\pm}$	$2\Gamma_5 + 2\Gamma_6 + 2\Gamma_7$
$D_{13/2}^{\pm}$	$3\Gamma_5 + 2\Gamma_6 + 2\Gamma_7$

(29) T_h. This group is the direct product of T and C_i ($T_h = T \times C_i$). Consider Figure 4. The additional operations in the group, besides those in T and C_i, are four $S_6 = C_3^{-1} \times I$ and four $S_6^{-1} = C_3 \times I$ operations and three reflections (σ_h) through the xy, yz, and zx planes.

Table 79. Character Table and Basis Functions for the Group T_h

T_h	E	$\bar{E}$	$3C_2$ $3\bar{C}_2$	$4C_3$	$4\bar{C}_3$	$4C_3^{-1}$	$4\bar{C}_3^{-1}$	I	$\bar{I}$	$3\sigma_h$ $3\bar{\sigma}_h$	$4S_6^{-1}$	$4\bar{S}_6^{-1}$	$4S_6$	$4\bar{S}_6$	Time Inv.	Bases
Γ_1^+	1	1	1	1	1	1	1	1	1	1	1	1	1	1	a	R
Γ_2^+	1	1	1	ω	ω	ω^2	ω^2	1	1	1	ω	ω	ω^2	ω^2	b	$1/\sqrt{2}\,(u+iv)$
Γ_3^+	1	1	1	ω^2	ω^2	ω	ω	1	1	1	ω^2	ω^2	ω	ω	b	$1/\sqrt{2}\,(u-iv)$
Γ_4^+	3	3	−1	0	0	0	0	3	3	−1	0	0	0	0	a	S_x, S_y, S_z
Γ_1^-	1	1	1	1	1	1	1	−1	−1	−1	−1	−1	−1	−1	a	xyz
Γ_2^-	1	1	1	ω	ω	ω^2	ω^2	−1	−1	−1	$-\omega$	$-\omega$	$-\omega^2$	$-\omega^2$	b	$\Gamma_2^+ \times \Gamma_1^-$
Γ_3^-	1	1	1	ω^2	ω^2	ω	ω	−1	−1	−1	$-\omega^2$	$-\omega^2$	$-\omega$	$-\omega$	b	$\Gamma_3^+ \times \Gamma_1^-$
Γ_4^-	3	3	−1	0	0	0	0	−3	−3	1	0	0	0	0	a	x,y,z
Γ_5^+	2	−2	0	1	−1	1	−1	2	−2	0	1	−1	1	−1	c	$\phi(1/2,-1/2)$, $\phi(1/2, 1/2)$
Γ_6^+	2	−2	0	ω	$-\omega$	ω^2	$-\omega^2$	2	−2	0	ω	$-\omega$	ω^2	$-\omega^2$	b	$\Gamma_5^+ \times \Gamma_2^+$
Γ_7^+	2	−2	0	ω^2	$-\omega^2$	ω	$-\omega$	2	−2	0	ω^2	$-\omega^2$	ω	$-\omega$	b	$(\Gamma_6^+)^*$
Γ_5^-	2	−2	0	1	−1	1	−1	−2	2	0	−1	1	−1	1	c	$\Gamma_5^+ \times \Gamma_1^-$
Γ_6^-	2	−2	0	ω	$-\omega$	ω^2	$-\omega^2$	−2	2	0	$-\omega$	ω	$-\omega^2$	ω^2	b	$\Gamma_5^+ \times \Gamma_2^-$
Γ_7^-	2	−2	0	ω^2	$-\omega^2$	ω	$-\omega$	−2	2	0	$-\omega^2$	ω^2	$-\omega$	ω	b	$(\Gamma_6^-)^*$

$$\omega = \exp(2\pi i/3)$$

$$u = 3z^2 - r^2, \qquad v = \sqrt{3}(x^2 - y^2)$$

Table 80. Compatibility Table

T_h	Γ_1^+	Γ_2^+	Γ_3^+	Γ_4^+	Γ_1^-	Γ_2^-	Γ_3^-	Γ_4^-	Γ_5^+	Γ_6^+	Γ_7^+	Γ_5^-	Γ_6^-	Γ_7^-
T	Γ_1	Γ_2	Γ_3	Γ_4	Γ_1	Γ_2	Γ_3	Γ_4	Γ_5	Γ_6	Γ_7	Γ_5	Γ_6	Γ_7
D_{2h}	Γ_1^+	Γ_1^+	Γ_1^+	$\Gamma_2^+ + \Gamma_3^+ + \Gamma_4^+$	Γ_1^-	Γ_1^-	Γ_1^-	$\Gamma_2^- + \Gamma_3^- + \Gamma_4^-$	Γ_5^+	Γ_5^+	Γ_5^+	Γ_5^-	Γ_5^-	Γ_5^-
C_{3i}:H(w)	Γ_1^+	Γ_2^+	Γ_3^+	$\Gamma_1^+ + \Gamma_2^+ + \Gamma_3^+$	Γ_1^-	Γ_2^-	Γ_3^-	$\Gamma_1^- + \Gamma_2^- + \Gamma_3^-$	$\Gamma_4^+ + \Gamma_5^+$	$\Gamma_4^+ + \Gamma_6^+$	$\Gamma_5^+ + \Gamma_6^+$	$\Gamma_4^- + \Gamma_5^-$	$\Gamma_4^- + \Gamma_6^-$	$\Gamma_5^- + \Gamma_6^-$
C_3:E(w)	Γ_1	Γ_2	Γ_3	$\Gamma_1 + \Gamma_2 + \Gamma_3$	Γ_1	Γ_2	Γ_3	$\Gamma_1 + \Gamma_2 + \Gamma_3$	$\Gamma_4 + \Gamma_5$	$\Gamma_4 + \Gamma_6$	$\Gamma_5 + \Gamma_6$	$\Gamma_4 + \Gamma_5$	$\Gamma_4 + \Gamma_6$	$\Gamma_5 + \Gamma_6$
C_{2v}:E(z)	Γ_1	Γ_1	Γ_1	$\Gamma_2 + \Gamma_3 + \Gamma_4$	Γ_3	Γ_3	Γ_3	$\Gamma_1 + \Gamma_2 + \Gamma_4$	Γ_5	Γ_5	Γ_5	Γ_5	Γ_5	Γ_5
C_{2h}:H(z)	Γ_1^+	Γ_1^+	Γ_1^+	$\Gamma_1^+ + 2\Gamma_2^+$	Γ_1^-	Γ_1^-	Γ_1^-	$\Gamma_1^- + 2\Gamma_2^-$	$\Gamma_3^+ + \Gamma_4^+$	$\Gamma_3^+ + \Gamma_4^+$	$\Gamma_3^+ + \Gamma_4^+$	$\Gamma_3^- + \Gamma_4^-$	$\Gamma_3^- + \Gamma_4^-$	$\Gamma_3^- + \Gamma_4^-$
C_s:E(v)	Γ_1	Γ_1	Γ_1	$\Gamma_1 + 2\Gamma_2$	Γ_2	Γ_2	Γ_2	$2\Gamma_1 + \Gamma_2$	$\Gamma_3 + \Gamma_4$	$\Gamma_3 + \Gamma_4$	$\Gamma_3 + \Gamma_4$	$\Gamma_3 + \Gamma_4$	$\Gamma_3 + \Gamma_4$	$\Gamma_3 + \Gamma_4$
C_i:H(v)	Γ_1^+	Γ_1^+	Γ_1^+	$3\Gamma_1^+$	Γ_1^-	Γ_1^-	Γ_1^-	$3\Gamma_1^-$	$2\Gamma_2^+$	$2\Gamma_2^+$	$2\Gamma_2^+$	$2\Gamma_2^-$	$2\Gamma_2^-$	$2\Gamma_2^-$

(30) O. This group consists of all proper rotations which send a cube into itself. There are four threefold axes O1, O2, O3, and O4 in Figure 4, three fourfold axes x, y and z and six twofold axes which pass through the origin O and bisect the twelve cube edges. There are therefore, in addition to the operations of T six operations C_4 and C_4^{-1} about the x, y, and z axes and six operations $C_{2'}$ about the six twofold axes.

(31) T_d. This group, isomorphic to O, consists of all those operations which send a regular tetrahedron into itself. In addition to the operations of T, we have three S_4 and three S_4^{-1} operations. They consist of rotations through ±90° about the axes x, y, and z each followed by a reflection in a plane perpendicular to the rotation axis, yz, xz, and xy respectively. There are also six reflections (σ_d) in the six planes which pass through the center of the cube and contain the six edges of the tetrahedron.

Table 81. Character Table and Basis Functions for the Groups O and T_d

O	E	$\bar{E}$	$8C_3$	$8\bar{C}_3$	$3C_2$ $3\bar{C}_2$	$6C_4$	$6\bar{C}_4$	$6C_2'$ $6\bar{C}_2'$			
T_d	E	$\bar{E}$	$8C_3$	$8\bar{C}_3$	$3C_2$ $3\bar{C}_2$	$6S_4$	$6\bar{S}_4$	$6\sigma_d$ $6\bar{\sigma}_d$	Time Inv.	Bases for O	Bases for T_d
Γ_1	1	1	1	1	1	1	1	1	a	R	R or xyz
Γ_2	1	1	1	1	1	−1	−1	−1	a	xyz	$S_xS_yS_z$
Γ_3	2	2	−1	−1	2	0	0	0	a	$(2z^2-x^2-y^2)$, $\sqrt{3}(x^2-y^2)$	$(2z^2-x^2-y^2)$, $\sqrt{3}(x^2-y^2)$
Γ_4	3	3	0	0	−1	1	1	−1	a	S_x, S_y, S_z	S_x, S_y, S_z
Γ_5	3	3	0	0	−1	−1	−1	1	a	yz, xz, xy	x, y, z
Γ_6	2	−2	1	−1	0	$\sqrt{2}$	$-\sqrt{2}$	0	c	$\phi(1/2,-1/2)$, $\phi(1/2, 1/2)$	$\phi(1/2,-1/2)$, $\phi(1/2, 1/2)$
Γ_7	2	−2	1	−1	0	$-\sqrt{2}$	$\sqrt{2}$	0	c	$\Gamma_6 \times \Gamma_2$	$\Gamma_6 \times \Gamma_2$
Γ_8	4	−4	−1	1	0	0	0	0	c	$\phi(3/2,-3/2)$, $\phi(3/2,-1/2)$ $\phi(3/2,1/2)$, $\phi(3/2,3/2)$	$\phi(3/2,-3/2)$, $\phi(3/2,-1/2)$ $\phi(3/2,1/2)$, $\phi(3/2,3/2)$

Table 82. Multiplication Table for the Groups O and T_d

Γ_1	Γ_2	Γ_3	Γ_4	Γ_5	Γ_6	Γ_7	Γ_8	
Γ_1	Γ_2	Γ_3	Γ_4	Γ_5	Γ_6	Γ_7	Γ_8	Γ_1
	Γ_1	Γ_3	Γ_5	Γ_4	Γ_7	Γ_6	Γ_8	Γ_2
		$\Gamma_1+\Gamma_2+\Gamma_3$	$\Gamma_4+\Gamma_5$	$\Gamma_4+\Gamma_5$	Γ_8	Γ_8	$\Gamma_6+\Gamma_7+\Gamma_8$	Γ_3
			$\Gamma_1+\Gamma_3+\Gamma_4+\Gamma_5$	$\Gamma_2+\Gamma_3+\Gamma_4+\Gamma_5$	$\Gamma_6+\Gamma_8$	$\Gamma_7+\Gamma_8$	$\Gamma_6+\Gamma_7+2\Gamma_8$	Γ_4
				$\Gamma_1+\Gamma_3+\Gamma_4+\Gamma_5$	$\Gamma_7+\Gamma_8$	$\Gamma_6+\Gamma_8$	$\Gamma_6+\Gamma_7+2\Gamma_8$	Γ_5
					$\Gamma_1+\Gamma_4$	$\Gamma_2+\Gamma_5$	$\Gamma_3+\Gamma_4+\Gamma_5$	Γ_6
						$\Gamma_1+\Gamma_4$	$\Gamma_3+\Gamma_4+\Gamma_5$	Γ_7
							$\Gamma_1+\Gamma_2+\Gamma_3$ $+2\Gamma_4+2\Gamma_5$	Γ_8

Table 83. Coupling Coefficients for the Groups O and T_d

u_1^3 transforms as $3z^2 - r^2$

u_2^3 transforms as $\sqrt{3}(x^2 - y^2)$

	$u_2v_1^3$	$u_2v_2^3$
ψ_1^3	0	1
ψ_2^3	−1	0

	$u_2v_{-1/2}^6$	$u_2v_{1/2}^6$
$\psi_{-1/2}^7$	1	0
$\psi_{1/2}^7$	0	1

	$u_2v_{-1/2}^7$	$u_2v_{1/2}^7$
$\psi_{-1/2}^6$	1	0
$\psi_{1/2}^6$	0	1

	$u_2v_x^4$	$u_2v_y^4$	$u_2v_z^4$
ψ_{yx}^5	1	0	0
ψ_{xz}^5	0	1	0
ψ_{yz}^5	0	0	1

	$u_2v_{yz}^5$	$u_2v_{xz}^5$	$u_2v_{xy}^5$
ψ_x^4	1	0	0
ψ_y^4	0	1	0
ψ_z^4	0	0	1

	$u_2v_{-3/2}^8$	$u_2v_{-1/2}^8$	$u_2v_{1/2}^8$	$u_2v_{3/2}^8$
$\psi_{-3/2}^8$	0	0	1	0
$\psi_{-1/2}^8$	0	0	0	−1
$\psi_{1/2}^8$	−1	0	0	0
$\psi_{3/2}^8$	0	1	0	0

	$u_1^3v_1^3$	$u_1^3v_2^3$	$u_2^3v_1^3$	$u_2^3v_2^3$
ψ_1	$1/\sqrt{2}$	0	0	$1/\sqrt{2}$
ψ_2	0	$1/\sqrt{2}$	$-1/\sqrt{2}$	0
ψ_1^3	$-1/\sqrt{2}$	0	0	$1/\sqrt{2}$
ψ_2^3	0	$1/\sqrt{2}$	$1/\sqrt{2}$	0

	$u_1^3v_x^4$	$u_1^3v_y^4$	$u_1^3v_z^4$	$u_2^3v_x^4$	$u_2^3v_y^4$	$u_2^3v_z^4$
ψ_x^4	−1/2	0	0	$\sqrt{3}/2$	0	0
ψ_y^4	0	−1/2	0	0	$-\sqrt{3}/2$	0
ψ_z^4	0	0	1	0	0	0
ψ_{yz}^5	$-\sqrt{3}/2$	0	0	−1/2	0	0
ψ_{xz}^5	0	$\sqrt{3}/2$	0	0	−1/2	0
ψ_{xy}^5	0	0	0	0	0	1

Table 83. Coupling Coefficients for the Groups O and T_d (Cont.)

	$u_1^3 v_{yz}^5$	$u_1^3 v_{xz}^5$	$u_1^3 v_{xy}^5$	$u_2^3 v_{yz}^5$	$u_2^3 v_{xz}^5$	$u_2^3 v_{xy}^5$
ψ_x^4	$-\sqrt{3}/2$	0	0	$-1/2$	0	0
ψ_y^4	0	$\sqrt{3}/2$	0	0	$-1/2$	0
ψ_z^4	0	0	0	0	0	1
ψ_{yz}^5	$-1/2$	0	0	$\sqrt{3}/2$	0	0
ψ_{xz}^5	0	$-1/2$	0	0	$-\sqrt{3}/2$	0
ψ_{xy}^5	0	0	1	0	0	0

	$u_1^3 v_{-1/2}^6$	$u_1^3 v_{1/2}^6$	$u_2^3 v_{-1/2}^6$	$u_2^3 v_{1/2}^6$
$\psi_{-3/2}^8$	0	0	0	1
$\psi_{-1/2}^8$	1	0	0	0
$\psi_{1/2}^8$	0	-1	0	0
$\psi_{3/2}^8$	0	0	-1	0

	$u_1^3 v_{-1/2}^7$	$u_1^3 v_{1/2}^7$	$u_2^3 v_{-1/2}^7$	$u_2^3 v_{1/2}^7$
$\psi_{-3/2}^8$	0	-1	0	0
$\psi_{-1/2}^8$	0	0	1	0
$\psi_{1/2}^8$	0	0	0	-1
$\psi_{3/2}^8$	1	0	0	0

	$u_1^3 v_{-3/2}^8$	$u_1^3 v_{-1/2}^8$	$u_1^3 v_{1/2}^8$	$u_1^3 v_{3/2}^8$	$u_2^3 v_{-3/2}^8$	$u_2^3 v_{-1/2}^8$	$u_2^3 v_{1/2}^8$	$u_2^3 v_{3/2}^8$
$\psi_{-1/2}^6$	0	$1/\sqrt{2}$	0	0	0	0	0	$1/\sqrt{2}$
$\psi_{1/2}^6$	0	0	$-1\sqrt{2}$	0	$-1/\sqrt{2}$	0	0	0
$\psi_{-1/2}^7$	0	0	0	$1/\sqrt{2}$	0	$-1/\sqrt{2}$	0	0
$\psi_{-3/2}^7$	$-1/\sqrt{2}$	0	0	0	0	0	$1/\sqrt{2}$	0
$\psi_{-3/2}^8$	$1/\sqrt{2}$	0	0	0	0	0	$1/\sqrt{2}$	0
$\psi_{-1/2}^8$	0	$-1/\sqrt{2}$	0	0	0	0	0	$1/\sqrt{2}$
$\psi_{1/2}^8$	0	0	$-1/\sqrt{2}$	0	$1/\sqrt{2}$	0	0	0
$\psi_{3/2}^8$	0	0	0	$1/\sqrt{2}$	0	$1/\sqrt{2}$	0	0

Table 83. Coupling Coefficients for the Groups O and T_d(Cont.)

	$u^4_x v^4_x$	$u^4_x v^4_y$	$u^4_x v^4_z$	$u^4_y v^4_x$	$u^4_y v^4_y$	$u^4_y v^4_z$	$u^4_z v^4_x$	$u^4_z v^4_y$	$u^4_z v^4_z$
ψ_1	$1/\sqrt{3}$	0	0	0	$1/\sqrt{3}$	0	0	0	$1/\sqrt{3}$
ψ^3_1	$-1/\sqrt{6}$	0	0	0	$-1/\sqrt{6}$	0	0	0	$\sqrt{2}/\sqrt{3}$
ψ^3_2	$1/\sqrt{2}$	0	0	0	$-1/\sqrt{2}$	0	0	0	0
ψ^4_x	0	0	0	0	0	$1/\sqrt{2}$	0	$-1/\sqrt{2}$	0
ψ^4_y	0	0	$-1/\sqrt{2}$	0	0	0	$1/\sqrt{2}$	0	0
ψ^4_z	0	$1/\sqrt{2}$	0	$-1/\sqrt{2}$	0	0	0	0	0
ψ^5_{yz}	0	0	0	0	0	$1/\sqrt{2}$	0	$1/\sqrt{2}$	0
ψ^5_{xz}	0	0	$1/\sqrt{2}$	0	0	0	$1/\sqrt{2}$	0	0
ψ^5_{xy}	0	$1/\sqrt{2}$	0	$1/\sqrt{2}$	0	0	0	0	0

	$u^4_x v^5_{yz}$	$u^4_x v^5_{xz}$	$u^4_x v^5_{xy}$	$u^4_y v^5_{yz}$	$u^4_y v^5_{xz}$	$u^4_y v^5_{xy}$	$u^4_z v^5_{yz}$	$u^4_z v^5_{xz}$	$u^4_z v^5_{xy}$
ψ_2	$1/\sqrt{3}$	0	0	0	$1/\sqrt{3}$	0	0	0	$1/\sqrt{3}$
ψ^3_1	$1/\sqrt{2}$	0	0	0	$-1/\sqrt{2}$	0	0	0	0
ψ^3_2	$1/\sqrt{6}$	0	0	0	$1/\sqrt{6}$	0	0	0	$-\sqrt{2}/\sqrt{3}$
ψ^4_x	0	0	0	0	0	$1/\sqrt{2}$	0	$1/\sqrt{2}$	0
ψ^4_y	0	0	$1/\sqrt{2}$	0	0	0	$1/\sqrt{2}$	0	0
ψ^4_z	0	$1/\sqrt{2}$	0	$1/\sqrt{2}$	0	0	0	0	0
ψ^5_{yz}	0	0	0	0	0	$1/\sqrt{2}$	0	$-1/\sqrt{2}$	0
ψ^5_{xz}	0	0	$-1/\sqrt{2}$	0	0	0	$1/\sqrt{2}$	0	0
ψ^5_{xy}	0	$1/\sqrt{2}$	0	$-1/\sqrt{2}$	0	0	0	0	0

	$u^4_x v^6_{-1/2}$	$u^4_x v^6_{1/2}$	$u^4_y v^6_{-1/2}$	$u^4_y v^6_{1/2}$	$u^4_z v^6_{-1/2}$	$u^4_z v^6_{1/2}$
$\psi^6_{-1/2}$	0	$-i/\sqrt{3}$	0	$-1/\sqrt{3}$	$i/\sqrt{3}$	0
$\psi^6_{1/2}$	$-i/\sqrt{3}$	0	$1/\sqrt{3}$	0	0	$-i/\sqrt{3}$
$\psi^8_{-3/2}$	$i/\sqrt{2}$	0	$1/\sqrt{2}$	0	0	0
$\psi^8_{-1/2}$	0	$i/\sqrt{6}$	0	$1/\sqrt{6}$	$i\sqrt{2}/\sqrt{3}$	0
$\psi^8_{1/2}$	$-i/\sqrt{6}$	0	$1/\sqrt{6}$	0	0	$i\sqrt{2}/\sqrt{3}$
$\psi^8_{3/2}$	0	$-i/\sqrt{2}$	0	$1/\sqrt{2}$	0	0

	$u^4_x v^7_{-1/2}$	$u^4_x v^7_{1/2}$	$u^4_y v^7_{-1/2}$	$u^4_y v^7_{1/2}$	$u^4_x v^7_{-1/2}$	$u^4_z v^7_{1/2}$
$\psi^7_{-1/2}$	0	$-i/\sqrt{3}$	0	$-1/\sqrt{3}$	$i/\sqrt{3}$	0
$\psi^7_{1/2}$	$-i/\sqrt{3}$	0	$1/\sqrt{3}$	0	0	$-i/\sqrt{3}$
$\psi^8_{-3/2}$	$-i/\sqrt{6}$	0	$1/\sqrt{6}$	0	0	$i\sqrt{2}/\sqrt{3}$
$\psi^8_{-1/2}$	0	$i/\sqrt{2}$	0	$-1/\sqrt{2}$	0	0
$\psi^8_{1/2}$	$-i/\sqrt{2}$	0	$-1/\sqrt{2}$	0	0	0
$\psi^8_{3/2}$	0	$i/\sqrt{6}$	0	$1/\sqrt{6}$	$i\sqrt{2}/\sqrt{3}$	0

Table 83. Coupling Coefficients for the Groups O and T_d (Cont.)

	$u_x^4 v_{-3/2}^8$	$u_x^4 v_{-1/2}^8$	$u_x^4 v_{1/2}^8$	$u_x^4 v_{3/2}^8$	$u_y^4 v_{-3/2}^8$	$u_y^4 v_{-1/2}^8$	$u_y^4 v_{1/2}^8$	$u_y^4 v_{3/2}^8$	$u_z^4 v_{-3/2}^8$	$u_z^4 v_{-1/2}^8$	$u_z^4 v_{1/2}^8$	$u_z^4 v_{3/2}^8$
$\psi_{-3/2}^8$	0	$i/\sqrt{5}$	0	0	0	$1/\sqrt{5}$	0	0	$-i\sqrt{3}/\sqrt{5}$	0	0	0
$\psi_{-1/2}^8$	$i/\sqrt{5}$	0	$i\sqrt{4}/\sqrt{15}$	0	$-1/\sqrt{5}$	0	$\sqrt{4}/\sqrt{15}$	0	0	$-i/\sqrt{15}$	0	0
$\psi_{1/2}^8$	0	$i\sqrt{4}/\sqrt{15}$	0	$i/\sqrt{5}$	0	$-\sqrt{4}/\sqrt{15}$	0	$1/\sqrt{5}$	0	0	$i/\sqrt{15}$	0
$\psi_{3/2}^8$	0	0	$i/\sqrt{5}$	0	0	0	$-1/\sqrt{5}$	0	0	0	0	$i\sqrt{3}/\sqrt{5}$
$\phi_{-3/2}^8$	0	$i/\sqrt{20}$	0	$-i\sqrt{5}/\sqrt{12}$	0	$1/\sqrt{20}$	0	$\sqrt{5}/\sqrt{12}$	$i/\sqrt{15}$	0	0	0
$\phi_{-1/2}^8$	$i/\sqrt{20}$	0	$-i\sqrt{3}/\sqrt{20}$	0	$-1/\sqrt{20}$	0	$-\sqrt{3}/\sqrt{20}$	0	0	$-i\sqrt{3}/\sqrt{5}$	0	0
$\phi_{1/2}^8$	0	$-i\sqrt{3}/\sqrt{20}$	0	$i/\sqrt{20}$	0	$\sqrt{3}/\sqrt{20}$	0	$1/\sqrt{20}$	0	0	$i\sqrt{3}/\sqrt{5}$	0
$\phi_{3/2}^8$	$-i\sqrt{5}/\sqrt{12}$	0	$i/\sqrt{20}$	0	$-\sqrt{5}/\sqrt{12}$	0	$-1/\sqrt{20}$	0	0	0	0	$-i/\sqrt{15}$
$\psi_{-1/2}^6$	$-i/2$	0	$i/\sqrt{12}$	0	$1/2$	0	$1/\sqrt{12}$	0	0	$-i/\sqrt{3}$	0	0
$\psi_{1/2}^6$	0	$-i/\sqrt{12}$	0	$i/2$	0	$1/\sqrt{12}$	0	$1/2$	0	0	$-i/\sqrt{3}$	0
$\psi_{-1/2}^7$	$i/\sqrt{12}$	0	$i/2$	0	$1/\sqrt{12}$	0	$-1/2$	0	0	0	0	$-i/\sqrt{3}$
$\psi_{1/2}^7$	0	$-i/2$	0	$-i/\sqrt{12}$	0	$-1/2$	0	$1/\sqrt{12}$	$-i/\sqrt{3}$	0	0	0

Table 83. Coupling Coefficients for the Groups O and T_d (Cont.)

	$u^5_{yz}v^5_{yz}$	$u^5_{yz}v^5_{xz}$	$u^5_{yz}v^5_{xy}$	$u^5_{xz}v^5_{yz}$	$u^5_{xz}v^5_{xz}$	$u^5_{xz}v^5_{xy}$	$u^5_{xy}v^5_{yz}$	$u^5_{xy}v^5_{xz}$	$u^5_{xy}v^5_{xy}$
ψ_1	$1/\sqrt{3}$	0	0	0	$1/\sqrt{3}$	0	0	0	$1/\sqrt{3}$
ψ^3_1	$-1/\sqrt{6}$	0	0	0	$-1/\sqrt{6}$	0	0	0	$\sqrt{2}/\sqrt{3}$
ψ^3_2	$1/\sqrt{2}$	0	0	0	$-1/\sqrt{2}$	0	0	0	0
ψ^4_x	0	0	0	0	0	$1/\sqrt{2}$	0	$-1/\sqrt{2}$	0
ψ^4_y	0	0	$-1/\sqrt{2}$	0	0	0	$1/\sqrt{2}$	0	0
ψ^4_z	0	$1/\sqrt{2}$	0	$-1/\sqrt{2}$	0	0	0	0	0
ψ^5_{yz}	0	0	0	0	0	$1/\sqrt{2}$	0	$1/\sqrt{2}$	0
ψ^5_{xz}	0	0	$1/\sqrt{2}$	0	0	0	$1/\sqrt{2}$	0	0
ψ^5_{xy}	0	$1/\sqrt{2}$	0	$1/\sqrt{2}$	0	0	0	0	0

	$u^5_{yz}v^6_{-1/2}$	$u^5_{yz}v^6_{1/2}$	$u^5_{xz}v^6_{-1/2}$	$u^5_{xz}v^6_{1/2}$	$u^5_{xy}v^6_{-1/2}$	$u^5_{xy}v^6_{1/2}$
$\psi^7_{-1/2}$	0	$-i/\sqrt{3}$	0	$-1/\sqrt{3}$	$i/\sqrt{3}$	0
$\psi^7_{1/2}$	$-i/\sqrt{3}$	0	$1/\sqrt{3}$	0	0	$-i/\sqrt{3}$
$\psi^8_{-3/2}$	$-i/\sqrt{6}$	0	$1/\sqrt{6}$	0	0	$i\sqrt{2}/\sqrt{3}$
$\psi^8_{-1/2}$	0	$i/\sqrt{2}$	0	$-1/\sqrt{2}$	0	0
$\psi^8_{1/2}$	$-i/\sqrt{2}$	0	$-1/\sqrt{2}$	0	0	0
$\psi^8_{3/2}$	0	$i/\sqrt{6}$	0	$1/\sqrt{6}$	$i\sqrt{2}/\sqrt{3}$	0

	$u^5_{yz}v^7_{-1/2}$	$u^5_{yz}v^7_{1/2}$	$u^5_{xz}v^7_{-1/2}$	$u^5_{xz}v^7_{1/2}$	$u^5_{xy}v^7_{-1/2}$	$u^5_{xy}v^7_{1/2}$
$\psi^6_{-1/2}$	0	$-i/\sqrt{3}$	0	$-1/\sqrt{3}$	$i/\sqrt{3}$	0
$\psi^6_{1/2}$	$-i/\sqrt{3}$	0	$1/\sqrt{3}$	0	0	$-i/\sqrt{3}$
$\psi^8_{-3/2}$	$i/\sqrt{2}$	0	$1/\sqrt{2}$	0	0	0
$\psi^8_{-1/2}$	0	$i/\sqrt{6}$	0	$1/\sqrt{6}$	$i\sqrt{2}/\sqrt{3}$	0
$\psi^8_{1/2}$	$-i/\sqrt{6}$	0	$1/\sqrt{6}$	0	0	$i\sqrt{2}/\sqrt{3}$
$\psi^8_{3/2}$	0	$-i/\sqrt{2}$	0	$1/\sqrt{2}$	0	0

Table 83. Coupling Coefficients for the Groups O and T_d (Cont.)

	$u^5_{yz}v^8_{-3/2}$	$u^5_{yz}v^8_{-1/2}$	$u^5_{yz}v^8_{1/2}$	$u^5_{yz}v^8_{3/2}$	$u^5_{xz}v^8_{-3/2}$	$u^5_{xz}v^8_{-1/2}$	$u^5_{xz}v^8_{1/2}$	$u^5_{xz}v^8_{3/2}$	$u^5_{xy}v^8_{-3/2}$	$u^5_{xy}v^8_{-1/2}$	$u^5_{xy}v^8_{1/2}$	$u^5_{xy}v^8_{3/2}$
$\psi^8_{-3/2}$	0	$i\sqrt{4}/\sqrt{15}$	0	$i/\sqrt{5}$	0	$-\sqrt{4}/\sqrt{15}$	0	$1/\sqrt{5}$	0	0	$i/\sqrt{15}$	0
$\psi^8_{-1/2}$	0	0	$-i/\sqrt{5}$	0	0	0	$1/\sqrt{5}$	0	0	0	0	$-i\sqrt{3}/\sqrt{5}$
$\psi^8_{1/2}$	0	$-i/\sqrt{5}$	0	0	0	$-1/\sqrt{5}$	0	0	$i\sqrt{3}/\sqrt{5}$	0	0	0
$\psi^8_{3/2}$	$i/\sqrt{5}$	0	$i\sqrt{4}/\sqrt{15}$	0	$-1/\sqrt{5}$	0	$\sqrt{4}/\sqrt{15}$	0	0	$-i/\sqrt{15}$	0	0
$\phi^8_{-3/2}$	0	$-i\sqrt{3}/\sqrt{20}$	0	$i/\sqrt{20}$	0	$\sqrt{3}/\sqrt{20}$	0	$1/\sqrt{20}$	0	0	$i\sqrt{3}/\sqrt{5}$	0
$\phi^8_{-1/2}$	$i\sqrt{5}/\sqrt{12}$	0	$-i/\sqrt{20}$	0	$\sqrt{5}/\sqrt{12}$	0	$1/\sqrt{20}$	0	0	0	0	$i/\sqrt{15}$
$\phi^8_{1/2}$	0	$-i/\sqrt{20}$	0	$i\sqrt{5}/\sqrt{12}$	0	$-1/\sqrt{20}$	0	$-\sqrt{5}/\sqrt{12}$	$-i/\sqrt{15}$	0	0	0
$\phi^8_{3/2}$	$i/\sqrt{20}$	0	$-i\sqrt{3}/\sqrt{20}$	0	$-1/\sqrt{20}$	0	$-\sqrt{3}/\sqrt{20}$	0	0	$-i\sqrt{3}/\sqrt{5}$	0	0
$\psi^7_{-1/2}$	$-i/2$	0	$i/\sqrt{12}$	0	$1/2$	0	$1/\sqrt{12}$	0	0	$-i/\sqrt{3}$	0	0
$\psi^7_{1/2}$	0	$-i/\sqrt{12}$	0	$i/2$	0	$1/\sqrt{12}$	0	$1/2$	0	0	$-i/\sqrt{3}$	0
$\psi^6_{-1/2}$	$i/\sqrt{12}$	0	$i/2$	0	$1/\sqrt{12}$	0	$-1/2$	0	0	0	0	$-i/\sqrt{3}$
$\psi^6_{1/2}$	0	$-i/2$	0	$-i/\sqrt{12}$	0	$-1/2$	0	$1/\sqrt{12}$	$-i/\sqrt{3}$	0	0	0

Table 83. Coupling Coefficients for the Groups O and T_d (Cont.)

	$u^6_{-1/2}v^6_{-1/2}$	$u^6_{-1/2}v^6_{1/2}$	$u^6_{1/2}v^6_{-1/2}$	$u^6_{1/2}v^6_{1/2}$
ψ_1	0	$-1/\sqrt{2}$	$1/\sqrt{2}$	0
ψ^4_x	$-i/\sqrt{2}$	0	0	$i/\sqrt{2}$
ψ^4_y	$1/\sqrt{2}$	0	0	$1/\sqrt{2}$
ψ^4_z	0	$-i/\sqrt{2}$	$-i/\sqrt{2}$	0

	$u^6_{-1/2}v^7_{-1/2}$	$u^6_{-1/2}v^7_{1/2}$	$u^6_{1/2}v^7_{-1/2}$	$u^6_{1/2}v^7_{1/2}$
ψ_2	0	$-1/\sqrt{2}$	$1/\sqrt{2}$	0
ψ^5_{yz}	$-i/\sqrt{2}$	0	0	$i/\sqrt{2}$
ψ^5_{xz}	$1/\sqrt{2}$	0	0	$1/\sqrt{2}$
ψ^5_{xy}	0	$-i/\sqrt{2}$	$-i/\sqrt{2}$	0

	$u^7_{-1/2}v^7_{-1/2}$	$u^7_{-1/2}v^7_{1/2}$	$u^7_{1/2}v^7_{-1/2}$	$u^7_{1/2}v^7_{1/2}$
ψ_1	0	$-1/\sqrt{2}$	$1/\sqrt{2}$	0
ψ^4_x	$-i/\sqrt{2}$	0	0	$i/\sqrt{2}$
ψ^4_y	$1/\sqrt{2}$	0	0	$1/\sqrt{2}$
ψ^4_z	0	$-i/\sqrt{2}$	$-i/\sqrt{2}$	0

Table 83. Coupling Coefficients for the Groups O and T_d (Cont.)

	$u^6_{-1/2}v^8_{-3/2}$	$u^6_{-1/2}v^8_{-1/2}$	$u^6_{-1/2}v^8_{1/2}$	$u^6_{-1/2}v^8_{3/2}$	$u^6_{1/2}v^8_{-3/2}$	$u^6_{1/2}v^8_{-1/2}$	$u^6_{1/2}v^8_{1/2}$	$u^6_{1/2}v^8_{3/2}$
ψ^4_x	0	$-i/\sqrt{8}$	0	$i\sqrt{3}/\sqrt{8}$	$i\sqrt{3}/\sqrt{8}$	0	$-i/\sqrt{8}$	0
ψ^4_y	0	$1/\sqrt{8}$	0	$\sqrt{3}/\sqrt{8}$	$-\sqrt{3}/\sqrt{8}$	0	$-1/\sqrt{8}$	0
ψ^4_z	0	0	$-i/\sqrt{2}$	0	0	$i/\sqrt{2}$	0	0
ψ^5_{yz}	0	$-i\sqrt{3}/\sqrt{8}$	0	$-i/\sqrt{8}$	$-i/\sqrt{8}$	0	$-i\sqrt{3}/\sqrt{8}$	0
ψ^5_{xz}	0	$-\sqrt{3}/\sqrt{8}$	0	$1/\sqrt{8}$	$-1/\sqrt{8}$	0	$\sqrt{3}/\sqrt{8}$	0
ψ^5_{xy}	$-i/\sqrt{2}$	0	0	0	0	0	0	$i/\sqrt{2}$
ψ^3_1	0	0	$-1/\sqrt{2}$	0	0	$-1/\sqrt{2}$	0	0
ψ^3_2	$-1/\sqrt{2}$	0	0	0	0	0	0	$-1/\sqrt{2}$

	$u^7_{-1/2}v^8_{-3/2}$	$u^7_{-1/2}v^8_{-1/2}$	$u^7_{-1/2}v^8_{1/2}$	$u^7_{-1/2}v^8_{3/2}$	$u^7_{1/2}v^8_{-3/2}$	$u^7_{1/2}v^8_{-1/2}$	$u^7_{1/2}v^8_{1/2}$	$u^7_{1/2}v^8_{3/2}$
ψ^4_x	0	$-i\sqrt{3}/\sqrt{8}$	0	$-i/\sqrt{8}$	$-i/\sqrt{8}$	0	$-i\sqrt{3}/\sqrt{8}$	0
ψ^4_y	0	$-\sqrt{3}/\sqrt{8}$	0	$1/\sqrt{8}$	$-1/\sqrt{8}$	0	$\sqrt{3}/\sqrt{8}$	0
ψ^4_z	$-i/\sqrt{2}$	0	0	0	0	0	0	$i/\sqrt{2}$
ψ^5_{yz}	0	$-i/\sqrt{8}$	0	$i\sqrt{3}/\sqrt{8}$	$i\sqrt{3}/\sqrt{8}$	0	$-i/\sqrt{8}$	0
ψ^5_{xz}	0	$1/\sqrt{8}$	0	$\sqrt{3}/\sqrt{8}$	$-\sqrt{3}/\sqrt{8}$	0	$-1/\sqrt{8}$	0
ψ^5_{xy}	0	0	$-i/\sqrt{2}$	0	0	$i/\sqrt{2}$	0	0
ψ^3_1	$-1/\sqrt{2}$	0	0	0	0	0	0	$-1/\sqrt{2}$
ψ^3_2	0	0	$1/\sqrt{2}$	0	0	$1/\sqrt{2}$	0	0

Table 83. Coupling Coefficients

	$u^8_{-3/2}v^8_{-3/2}$	$u^8_{-3/2}v^8_{-1/2}$	$u^8_{-3/2}v^8_{1/2}$	$u^8_{-3/2}v^8_{3/2}$	$u^8_{-1/2}v^8_{-3/2}$	$u^8_{-1/2}v^8_{-1/2}$	$u^8_{-1/2}v^8_{1/2}$	$u^8_{-1/2}v^8_{3/2}$
ψ_1	0	0	0	$-1/2$	0	0	$1/2$	0
ψ_2	0	$-1/2$	0	0	$-1/2$	0	0	0
ψ_1^3	0	0	0	$1/2$	0	0	$1/2$	0
ψ_2^3	0	$1/2$	0	0	$-1/2$	0	0	0
ψ_x^4	0	0	$-i\sqrt{3}/\sqrt{20}$	0	0	$i/\sqrt{5}$	0	$i\sqrt{3}/\sqrt{20}$
ψ_y^4	0	0	$\sqrt{3}/\sqrt{20}$	0	0	$-1/\sqrt{5}$	0	$\sqrt{3}/\sqrt{20}$
ψ_z^4	0	0	0	$-i3/\sqrt{20}$	0	0	$i/\sqrt{20}$	0
ϕ_x^4	$-i\sqrt{5}/4$	0	$i\sqrt{3}/\sqrt{80}$	0	0	$i3/\sqrt{80}$	0	$-i\sqrt{3}/\sqrt{80}$
ϕ_y^4	$-\sqrt{5}/4$	0	$-\sqrt{3}/\sqrt{80}$	0	0	$-3/\sqrt{80}$	0	$-\sqrt{3}/\sqrt{80}$
ϕ_z^4	0	0	0	$-i/\sqrt{20}$	0	0	$-i3/\sqrt{20}$	0
ψ_{yz}^5	0	0	$i/2$	0	0	0	0	$i/2$
ψ_{xz}^5	0	0	$1/2$	0	0	0	0	$-1/2$
ψ_{xy}^5	0	$i/2$	0	0	$-i/2$	0	0	0
ϕ_{yz}^5	$-i\sqrt{3}/4$	0	$-i/4$	0	0	$-i\sqrt{3}/4$	0	$i/4$
ϕ_{xz}^5	$\sqrt{3}/4$	0	$-1/4$	0	0	$-\sqrt{3}/4$	0	$-1/4$
ϕ_{xy}^5	0	$i/2$	0	0	$i/2$	0	0	0

for the Groups O and T_d (Cont.)

	$u^8_{1/2}v^8_{-3/2}$	$u^8_{1/2}v^8_{-1/2}$	$u^8_{1/2}v^8_{1/2}$	$u^8_{1/2}v^8_{3/2}$	$u^8_{3/2}v^8_{-3/2}$	$u^8_{3/2}v^8_{-1/2}$	$u^8_{3/2}v^8_{1/2}$	$u^8_{3/2}v^8_{3/2}$
ψ_1	0	$-1/2$	0	0	$1/2$	0	0	0
ψ_2	0	0	0	$1/2$	0	0	$1/2$	0
ψ^3_1	0	$-1/2$	0	0	$-1/2$	0	0	0
ψ^3_2	0	0	0	$1/2$	0	0	$-1/2$	0
ψ^4_x	$-i\sqrt{3}/\sqrt{20}$	0	$-i/\sqrt{5}$	0	0	$i\sqrt{3}/\sqrt{20}$	0	0
ψ^4_y	$\sqrt{3}/\sqrt{20}$	0	$-1/\sqrt{5}$	0	0	$\sqrt{3}/\sqrt{20}$	0	0
ψ^4_z	0	$i/\sqrt{20}$	0	0	$-i3/\sqrt{20}$	0	0	0
ϕ^4_x	$i\sqrt{3}/\sqrt{80}$	0	$-i3/\sqrt{80}$	0	0	$-i\sqrt{3}/\sqrt{80}$	0	$i\sqrt{5}/4$
ϕ^4_y	$-\sqrt{3}/\sqrt{80}$	0	$-3/\sqrt{80}$	0	0	$-\sqrt{3}/\sqrt{80}$	0	$-\sqrt{5}/4$
ϕ^4_z	0	$-i3/\sqrt{20}$	0	0	$-i/\sqrt{20}$	0	0	0
ψ^5_{yz}	$-i/2$	0	0	0	0	$-i/2$	0	0
ψ^5_{xz}	$-1/2$	0	0	0	0	$1/2$	0	0
ψ^5_{xy}	0	0	0	$-i/2$	0	0	$i/2$	0
ϕ^5_{yz}	$-i/4$	0	$i\sqrt{3}/4$	0	0	$i/4$	0	$i\sqrt{3}/4$
ϕ^5_{xz}	$-1/4$	0	$-\sqrt{3}/4$	0	0	$-1/4$	0	$\sqrt{3}/4$
ϕ^5_{xy}	0	0	0	$i/2$	0	0	$i/2$	0

Table 84. Compatibility Table

T_d	O	Γ_1	Γ_2	Γ_3	Γ_4	Γ_5	Γ_6	Γ_7	Γ_8
T	T	Γ_1	Γ_1	$\Gamma_2 + \Gamma_3$	Γ_4	Γ_4	Γ_5	Γ_5	$\Gamma_6 + \Gamma_7$
D_{2d}	D_4	Γ_1	Γ_3	$\Gamma_1 + \Gamma_3$	$\Gamma_2 + \Gamma_5$	$\Gamma_4 + \Gamma_5$	Γ_6	Γ_7	$\Gamma_6 + \Gamma_7$
C_{3v} : E(w)	D_3	Γ_1	Γ_2	Γ_3	$\Gamma_2 + \Gamma_3$	$\Gamma_1 + \Gamma_3$	Γ_4	Γ_4	$\Gamma_4 + \Gamma_5 + \Gamma_6$
S_4 : H(z)	C_4 : H(z) : E(z)	Γ_1	Γ_2	$\Gamma_1 + \Gamma_2$	$\Gamma_1 + \Gamma_3 + \Gamma_4$	$\Gamma_2 + \Gamma_3 + \Gamma_4$	$\Gamma_5 + \Gamma_6$	$\Gamma_7 + \Gamma_8$	$\Gamma_5 + \Gamma_6 + \Gamma_7 + \Gamma_8$
C_3 : H(w)	C_3 : E(w) : H(w)	Γ_1	Γ_1	$\Gamma_2 + \Gamma_3$	$\Gamma_1 + \Gamma_2 + \Gamma_3$	$\Gamma_1 + \Gamma_2 + \Gamma_3$	$\Gamma_4 + \Gamma_5$	$\Gamma_4 + \Gamma_5$	$\Gamma_4 + \Gamma_5 + 2\Gamma_6$
C_{2v} : E(z)		Γ_1	Γ_3	$\Gamma_1 + \Gamma_3$	$\Gamma_2 + \Gamma_3 + \Gamma_4$	$\Gamma_1 + \Gamma_2 + \Gamma_4$	Γ_5	Γ_5	$2\Gamma_5$
C_s : E(v) : H(v)	C_2 : E(v) : H(v)	Γ_1	Γ_2	$\Gamma_1 + \Gamma_2$	$\Gamma_1 + 2\Gamma_2$	$2\Gamma_1 + \Gamma_2$	$\Gamma_3 + \Gamma_4$	$\Gamma_3 + \Gamma_4$	$2\Gamma_3 + 2\Gamma_4$

Table 85. Full Rotation Group Compatibility Table for the Group O

$D_0^{\pm}$	Γ_1
$D_1^{\pm}$	Γ_4
$D_2^{\pm}$	$\Gamma_3 + \Gamma_5$
$D_3^{\pm}$	$\Gamma_2 + \Gamma_4 + \Gamma_5$
$D_4^{\pm}$	$\Gamma_1 + \Gamma_3 + \Gamma_4 + \Gamma_5$
$D_5^{\pm}$	$\Gamma_3 + 2\Gamma_4 + \Gamma_5$
$D_6^{\pm}$	$\Gamma_1 + \Gamma_2 + \Gamma_3 + \Gamma_4 + 2\Gamma_5$
$D_{1/2}^{\pm}$	Γ_6
$D_{3/2}^{\pm}$	Γ_8
$D_{5/2}^{\pm}$	$\Gamma_7 + \Gamma_8$
$D_{7/2}^{\pm}$	$\Gamma_6 + \Gamma_7 + \Gamma_8$
$D_{9/2}^{\pm}$	$\Gamma_6 + 2\Gamma_8$
$D_{11/2}^{\pm}$	$\Gamma_6 + \Gamma_7 + 2\Gamma_8$
$D_{13/2}^{\pm}$	$\Gamma_6 + 2\Gamma_7 + 2\Gamma_8$

Table 86. Full Rotation Group Compatibility Table for the Group T_d

D_0^{+}	Γ_1	D_0^{-}	Γ_2
D_1^{+}	Γ_4	D_1^{-}	Γ_5
D_2^{+}	$\Gamma_3 + \Gamma_5$	D_2^{-}	$\Gamma_3 + \Gamma_4$
D_3^{+}	$\Gamma_2 + \Gamma_4 + \Gamma_5$	D_3^{-}	$\Gamma_1 + \Gamma_4 + \Gamma_5$
D_4^{+}	$\Gamma_1 + \Gamma_3 + \Gamma_4 + \Gamma_5$	D_4^{-}	$\Gamma_2 + \Gamma_3 + \Gamma_4 + \Gamma_5$
D_5^{+}	$\Gamma_3 + 2\Gamma_4 + \Gamma_5$	D_5^{-}	$\Gamma_3 + \Gamma_4 + 2\Gamma_5$
D_6^{+}	$\Gamma_1 + \Gamma_2 + \Gamma_3 + \Gamma_4 + 2\Gamma_5$	D_6^{-}	$\Gamma_1 + \Gamma_2 + \Gamma_3 + 2\Gamma_4 + \Gamma_5$
$D_{1/2}^{+}$	Γ_6	$D_{1/2}^{-}$	Γ_7
$D_{3/2}^{+}$	Γ_8	$D_{3/2}^{-}$	Γ_8
$D_{5/2}^{+}$	$\Gamma_7 + \Gamma_8$	$D_{5/2}^{-}$	$\Gamma_6 + \Gamma_8$
$D_{7/2}^{+}$	$\Gamma_6 + \Gamma_7 + \Gamma_8$	$D_{7/2}^{-}$	$\Gamma_6 + \Gamma_7 + \Gamma_8$
$D_{9/2}^{+}$	$\Gamma_6 + 2\Gamma_8$	$D_{9/2}^{-}$	$\Gamma_7 + 2\Gamma_8$
$D_{11/2}^{+}$	$\Gamma_6 + \Gamma_7 + 2\Gamma_8$	$D_{11/2}^{-}$	$\Gamma_6 + \Gamma_7 + 2\Gamma_8$
$D_{13/2}^{+}$	$\Gamma_6 + 2\Gamma_7 + 2\Gamma_8$	$D_{13/2}^{-}$	$2\Gamma_6 + \Gamma_7 + 2\Gamma_8$

(32) O_h. This group is the direct product of O and C_1 ($O_h = O \times C_1$) and consists of all operations which send a cube into itself. It may also be considered as the direct product of T_d and C_i and contains all the operations of O, T_d, and T_h.

Table 87. Character Table and Basis Functions for the Group O_h

O_h	E	$\bar{E}$	$8C_3$	$8\bar{C}_3$	$3C_2$ $3\bar{C}_2$	$6C_4$	$6\bar{C}_4$	$6C_2'$ $6\bar{C}_2'$	I	$\bar{I}$	$8S_6$	$8\bar{S}_6$	$3\sigma_h$ $3\bar{\sigma}_h$	$6S_4$	$6\bar{S}_4$	$6\sigma_d$ $6\bar{\sigma}_d$	Time Inv.	Bases
Γ_1^+	1	1	1	1	1	1	1	1	1	1	1	1	1	1	1	1	a	R
Γ_2^+	1	1	1	1	1	−1	−1	−1	1	1	1	1	1	−1	−1	−1	a	$(x^2-y^2)(y^2-z^2)(z^2-x^2)$
Γ_3^+	2	2	−1	−1	2	0	0	0	2	2	−1	−1	2	0	0	0	a	$(2z^2-x^2-y^2), \sqrt{3}(x^2-y^2)$
Γ_4^+	3	3	0	0	−1	1	1	−1	3	3	0	0	−1	1	1	−1	a	S_x, S_y, S_z
Γ_5^+	3	3	0	0	−1	−1	−1	1	3	3	0	0	−1	−1	−1	1	a	yz, xz, xy
Γ_1^-	1	1	1	1	1	1	1	1	−1	−1	−1	−1	−1	−1	−1	−1	a	$\Gamma_2^- \times \Gamma_2^+$
Γ_2^-	1	1	1	1	1	−1	−1	−1	−1	−1	−1	−1	−1	1	1	1	a	xyz
Γ_3^-	2	2	−1	−1	2	0	0	0	−2	−2	1	1	−2	0	0	0	a	$\Gamma_3^+ \times \Gamma_2^-$
Γ_4^-	3	3	0	0	−1	1	1	−1	−3	−3	0	0	1	−1	−1	1	a	x, y, z
Γ_5^-	3	3	0	0	−1	−1	−1	1	−3	−3	0	0	1	1	1	−1	a	$\Gamma_5^+ \times \Gamma_1^-$
Γ_6^+	2	−2	1	−1	0	$\sqrt{2}$	$-\sqrt{2}$	0	2	−2	1	−1	0	$\sqrt{2}$	$-\sqrt{2}$	0	c	$\phi(1/2,-1/2), \phi(1/2,1/2)$
Γ_7^+	2	−2	1	−1	0	$-\sqrt{2}$	$\sqrt{2}$	0	2	−2	1	−1	0	$-\sqrt{2}$	$\sqrt{2}$	0	c	$\Gamma_6^+ \times \Gamma_2^+$
Γ_8^+	4	−4	−1	1	0	0	0	0	4	−4	−1	1	0	0	0	0	c	$\phi(3/2,-3/2), \phi(3/2,-1/2)$ $\phi(3/2,1/2), \phi(3/2,3/2)$
Γ_6^-	2	−2	1	−1	0	$\sqrt{2}$	$-\sqrt{2}$	0	−2	2	−1	1	0	$-\sqrt{2}$	$\sqrt{2}$	0	c	$\Gamma_6^+ \times \Gamma_1^-$
Γ_7^-	2	−2	1	−1	0	$-\sqrt{2}$	$\sqrt{2}$	0	−2	2	−1	1	0	$\sqrt{2}$	$-\sqrt{2}$	0	c	$\Gamma_6^+ \times \Gamma_2^-$
Γ_8^-	4	−4	−1	1	0	0	0	0	−4	4	1	−1	0	0	0	0	c	$\Gamma_8^+ \times \Gamma_1^-$

Table 88. Compatibility Table

O_h	Γ_1^+	Γ_2^+	Γ_3^+	Γ_4^+	Γ_5^+	Γ_1^-	Γ_2^-	Γ_3^-	Γ_4^-	Γ_5^-
O	Γ_1	Γ_2	Γ_3	Γ_4	Γ_5	Γ_1	Γ_2	Γ_3	Γ_4	Γ_5
T_d	Γ_1	Γ_2	Γ_3	Γ_4	Γ_5	Γ_2	Γ_1	Γ_3	Γ_5	Γ_4
T_h	Γ_1^+	Γ_1^+	$\Gamma_2^+ + \Gamma_3^+$	Γ_4^+	Γ_4^+	Γ_1^-	Γ_1^-	$\Gamma_2^- + \Gamma_3^-$	Γ_4^-	Γ_4^-
D_{4h}	Γ_1^+	Γ_3^+	$\Gamma_1^+ + \Gamma_3^+$	$\Gamma_2^+ + \Gamma_5^+$	$\Gamma_4^+ + \Gamma_5^+$	Γ_1^-	Γ_3^-	$\Gamma_1^- + \Gamma_3^-$	$\Gamma_2^- + \Gamma_5^-$	$\Gamma_4^- + \Gamma_5^-$
D_{3d}	Γ_1^+	Γ_2^+	Γ_3^+	$\Gamma_2^+ + \Gamma_3^+$	$\Gamma_1^+ + \Gamma_3^+$	Γ_1^-	Γ_2^-	Γ_3^-	$\Gamma_2^- + \Gamma_3^-$	$\Gamma_1^- + \Gamma_3^-$
C_{4h} : H(z)	Γ_1^+	Γ_2^+	$\Gamma_1^+ + \Gamma_2^+$	$\Gamma_1^+ + \Gamma_3^+ + \Gamma_4^+$	$\Gamma_2^+ + \Gamma_3^+ + \Gamma_4^+$	Γ_1^-	Γ_2^-	$\Gamma_1^- + \Gamma_2^-$	$\Gamma_1^- + \Gamma_3^- + \Gamma_4^-$	$\Gamma_2^- + \Gamma_3^- + \Gamma$
C_{2h} : H(v)	Γ_1^+	Γ_2^+	$\Gamma_1^+ + \Gamma_2^+$	$2\Gamma_2^+ + \Gamma_1^+$	$2\Gamma_1 + \Gamma_2^+$	Γ_1^-	Γ_2^-	$\Gamma_1^- + \Gamma_2^-$	$2\Gamma_2^- + \Gamma_1^-$	$2\Gamma_1^- + \Gamma_2^-$
C_{3i} : H(w)	Γ_1^+	Γ_1^+	$\Gamma_2^+ + \Gamma_3^+$	$\Gamma_1^+ + \Gamma_2^+ + \Gamma_3^+$	$\Gamma_1^+ + \Gamma_2^+ + \Gamma_3^+$	Γ_1^-	Γ_1^-	$\Gamma_2^- + \Gamma_3^-$	$\Gamma_1^- + \Gamma_2^- + \Gamma_3^-$	$\Gamma_1^- + \Gamma_2^- + \Gamma$
C_{4v} : E(z)	Γ_1	Γ_3	$\Gamma_1 + \Gamma_3$	$\Gamma_2 + \Gamma_5$	$\Gamma_4 + \Gamma_5$	Γ_2	Γ_4	$\Gamma_2 + \Gamma_4$	$\Gamma_1 + \Gamma_5$	$\Gamma_3 + \Gamma_5$
C_{2v} : E(v)	Γ_1	Γ_2	$\Gamma_1 + \Gamma_2$	$\Gamma_2 + \Gamma_3 + \Gamma_4$	$\Gamma_1 + \Gamma_3 + \Gamma_4$	Γ_3	Γ_4	$\Gamma_3 + \Gamma_4$	$\Gamma_1 + \Gamma_2 + \Gamma_4$	$\Gamma_1 + \Gamma_2 + \Gamma$
C_{3v} : E(w)	Γ_1	Γ_2	Γ_3	$\Gamma_2 + \Gamma_3$	$\Gamma_1 + \Gamma_3$	Γ_2	Γ_1	Γ_3	$\Gamma_1 + \Gamma_3$	$\Gamma_2 + \Gamma_3$

O_h	Γ_6^+	Γ_7^+	Γ_8^+	Γ_6^-	Γ_7^-	Γ_8^-
O	Γ_6	Γ_7	Γ_8	Γ_6	Γ_7	Γ_8
T_d	Γ_6	Γ_7	Γ_8	Γ_7	Γ_6	Γ_8
T_h	Γ_5^+	Γ_5^+	$\Gamma_6^+ + \Gamma_7^+$	Γ_5^-	Γ_5^-	$\Gamma_6^- + \Gamma_7^-$
D_{4h}	Γ_6^+	Γ_7^+	$\Gamma_6^+ + \Gamma_7^+$	Γ_6^-	Γ_7^-	$\Gamma_6^- + \Gamma_7^-$
D_{3d}	Γ_4^+	Γ_4^+	$\Gamma_4^+ + \Gamma_5^+ + \Gamma_6^+$	Γ_4^-	Γ_4^-	$\Gamma_4^- + \Gamma_5^- + \Gamma_6^-$
C_{4h} : H(z)	$\Gamma_5^+ + \Gamma_6^+$	$\Gamma_7^+ + \Gamma_8^+$	$\Gamma_5^+ + \Gamma_6^+ + \Gamma_7^+ + \Gamma_8^+$	$\Gamma_5^- + \Gamma_6^-$	$\Gamma_7^- + \Gamma_8^-$	$\Gamma_5^- + \Gamma_6^- + \Gamma_7^- + \Gamma_8^-$
C_{2h} : H(v)	$\Gamma_3^+ + \Gamma_4^+$	$\Gamma_3^+ + \Gamma_4^+$	$2\Gamma_3^+ + 2\Gamma_4^+$	$\Gamma_3^- + \Gamma_4^-$	$\Gamma_3^- + \Gamma_4^-$	$2\Gamma_3^- + 2\Gamma_4^-$
C_{3i} : H(w)	$\Gamma_4^+ + \Gamma_5^+$	$\Gamma_4^+ + \Gamma_5^+$	$\Gamma_4^+ + \Gamma_5^+ + 2\Gamma_6^+$	$\Gamma_4^- + \Gamma_5^-$	$\Gamma_4^- + \Gamma_5^-$	$\Gamma_4^- + \Gamma_5^- + 2\Gamma_6^-$
C_{4v} : E(z)	Γ_6	Γ_7	$\Gamma_6 + \Gamma_7$	Γ_6	Γ_7	$\Gamma_6 + \Gamma_7$
C_{2v} : E(v)	Γ_5	Γ_5	$2\Gamma_5$	Γ_5	Γ_5	$2\Gamma_5$
C_{3v} : E(w)	Γ_4	Γ_4	$\Gamma_4 + \Gamma_5 + \Gamma_6$	Γ_4	Γ_4	$\Gamma_4 + \Gamma_5 + \Gamma_6$